Joseph Needham

# Joseph Needham

## 20th-Century Renaissance Man

by Maurice Goldsmith

*Profiles*

UNESCO PUBLISHING

Published in 1995 by the United Nations Educational,
Scientific and Cultural Organization,
7, place de Fontenoy, 75352 Paris 07 SP (France)

Layout: Jean-Francis Chériez
Composition: Éditions du Mouflon, 94270 Le Kremlin-Bicêtre (France)
Printed by Imprimerie Jouve, Paris

ISBN 92-3-103192-9

# Foreword

The following poem, from the collection entitled *Patterns: Poems by Federico Mayor* (London, Forest Books, 1994), was selected by the Director-General of UNESCO, Federico Mayor, as a testimony to Joseph Needham, at the author's request.

*They filled my hands with flowers*
*and my neck with garlands.*
*Oh, no, don't bow down to me.*
*Your welcome makes me blush*
*and reminds me*
*that you have nothing to thank me for!*
*. . . Later we climb*
*to the Swayambhnatu Temples*
*where the mystic*
*is our equal,*
*preceded as we are by banners,*
*trumpets, ringing of bells*
*and prayers.*
*You, young lama,*
*and I*
*together have felt*
*the same light and dark.*
*Beyond,*
*looms invisible,*
*yet so certain,*
*Everest.*

*In this land high and steep*
*people are born, survive*
*and die*
*unnoticed,*
*indescribable.*
*While we go to mass on Sundays*
*and holy days,*
*hold raffles and draws*
*in guise of charity*
*and regulate everything,*
*even love . . .*
*you – and you, young lama –*
*think we have understood*
*nothing.*
*Neither the manger, nor the cross,*
*nor the message of Buddha, your lord.*
*When I leave you among the offerings*
*and lamps of Swayambhu*
*I know that*
*we have still not learned*
*to love.*

Katmandu, 17 December 1988

# Contents

*To Anna, who understands*

*Nature from growing trees we best discern*
*And man's estate from social order learn.*

Translation into Popian Couplet
by Joseph Needham
of *Rjêng tao ming chêng*
*Ti tao ming shu*
Verse by Dr Shih Seng-Han of Wuhan University,
written in June 1943.

# List of illustrations

# Acknowledgements

My thanks to Joseph Needham, who gave me every encouragement; his two personal assistants Corinne Lucheux and Tracey Humphries for their never-failing support; various members of the Needham Research Institute (including the Director, Professor Ho Peng Yoke, and Deputy Director Dr Christopher Cullen; Angela King; John Moffet, the librarian, and his colleagues; Stanley Besh and Duncan Manson); Augusta Morton, my invaluable assistant, whose patience is astonishing; Dr Diana Holdright and her mother Angela Ford, relatives of Needham; Alan Mackay, David Barron and David Goldsmith for their comments on the manuscript; my wife Anna, for putting up once again with my author's 'nerves'.

I would also like to thank the United Kingdom Economic and Social Research Council for financial support, particularly former Chief Executive Professor Howard Newby, his secretary, Sarah Parker, and the current Chief Executive, Professor Ron Amann; Federico Mayor, Director-General of UNESCO, and his assistant Tom Forstenzer; the late Dr Wang Ling and his wife Ruth; Dr Christopher Brooke, historian of Gonville and Caius College; the librarians of Cambridge University, the Royal Society and the Athenaeum Club; Margaret Quass; Frank MacManus; and Professor A. Rahman, historian of Indian science. My thanks also to the Cambridge University Press for granting permission to include extracts from *Science and Civilisation in China* and to my indexer, Joe Britton.

Colin Ronan, whose sudden death in June 1995 is a great loss, had provided some valuable comments: in particular, to adopt the Pinyin transliteration system throughout the book. He had used this in his *Shorter Science and Civilisation in China,* his ongoing abridgement of Needham's original text.

I have used Needham's own modification of the Wade-Giles transliteration system for Chinese words in my biography, and the ordinary (unmodified) Wade-Giles system for Wang Ling's 'Reminiscences'. I have added the Pinyin transliteration in parentheses following the first mention.

# The author
# to Joseph Needham

You are a charming and highly intelligent person, one whose company I have enjoyed over many years. I am delighted that you should have asked me to write your authorized biography.

This work, published by UNESCO to coincide with the celebration of its fiftieth anniversary, as you were largely responsible for the 'S' in UNESCO, is not the major full-length study that you merit and on which I am still working. This is merely an introduction to introduce you to a wider audience.

But it has not been an easy task. Your challenge to me has been particularly heavy, quite different from the challenge I met when writing of the life of your intimate friend, Desmond Bernal,[1] and that of Frédéric Joliot-Curie.[2]

They were in a sense 'more simple and direct' than you. I find you oh-so-much-more complicated. Of course, I was fortunate that for three years I was able to visit you regularly in Cambridge and have many hours of discussion, all recorded.

What, then, emerges? Clearly, unlike most people, at a very early age, certainly before your teens, you became responsible, as Albert Camus put it, for your 'face'. I describe that in my opening chapter and then go on to clarify how your 'face' has been further shaped by what life has done to you. As a result you emerge as an individual, rare in any period, who through a

1.  Maurice Goldsmith, *Sage, A Life of J. D. Bernal*, London, Heinemann, 1980.
2.  Maurice Goldsmith, *Frédéric Joliot-Curie*, London, Lawrence & Wishart, 1976.

deeply held unitary world-view has developed concepts that have helped to reshape our social, political and theological views and behaviour.

From childhood onwards you have seen yourself as a 'bridge-builder': searching always, as you say, 'for a reconciliation, for a union of things separated'. You have been successful in storming us in four main fields: morphology and biochemistry; science and religion; socialism and social responsibility; and, above all, in China and the Western world.

In that last field, the die was cast in 1937 when four Chinese post-graduate students came to Cambridge to do research in biochemistry and you fell in love with China. You began to orchestrate a new reality in seeking to provide an answer to the key question of why Chinese science never developed as did Western science at the Renaissance. That was when your second half-life began. You became the historian of Chinese science and civilization, and in so doing, you changed historiography world-wide.

You have allied polymath learning with a powerful imagination, a mind open to all forms of cultural experience. This Confucian analect is relevant: 'For him who respects the dignity of man, and practises what love and courtesy require, for him all men within the four seas are brothers.'

And yet, and yet – I found your deeply held religious beliefs difficult to square with your activity as a scientist. As you point out, that is my problem not yours. And, once again, you have acted as a 'bridge-builder.' As American cosmologist

Frank J. Tipler states: 'It is time scientists reconsider the God hypothesis. . . . The time has come to absorb theology into physics, to make Heaven as real as an electron.'[3]

For you holiness is not just a belief, but an experience as vivid as sexual pleasure or hunger,[4] and the Kingdom of God on Earth – that is, a world co-operative community – is a dream that has been, and still is, with you each day of your long life.

I see you as an authentically great man. This biography is an introduction to explain why I do.

Maurice Goldsmith

P.S. Since I wrote the above, I have experienced the sadness of your death on the evening of Friday, 24 March 1995. What has intrigued me is the extensive space which the broadsheet dailies, in particular, have devoted to your obituary.

With your death you have crossed what J. W. Lambert called 'the mysterious barrier separating the admired from the famous', that is, you have become a household name.

I can hear you muttering, as you did when the Queen presented you with the Companion of Honour in 1992, 'About time, too.'

3. Frank J. Tipler, *The Physics of Immortality*, p. xv, London, Macmillan, 1995.
4. I quote from Marge Piercy's magnificent book, *Body of Glass*, p. 245, London, Michael Joseph, 1992.

# 1

# Parents and son

*I stood in Venice, on the Bridge of Sighs,*
*A palace and a prison on each hand.*

Byron, *Childe Harold*, IV.1

As the shouting dies and the brutality of crashing crockery ceases, the lonely five-year-old – Noel Joseph Terence Montgomery Needham – tiptoes cautiously from his top-floor playroom. He hesitates, then continues down to the 'battlefield'. He knows what will happen when he opens the door. His mother will say, 'Terence, darling', and his father will stand silent, mouthing 'Noel', his preferred name for the boy, born on 9 December 1900, as the Christmastime atmosphere was beginning to be felt.

His mother insisted on 'Terence', because it was Irish, symbolizing her homeland. Why then did he come to be called 'Joseph'? At about the age of 11 he discovered that Joseph was a name traditional in his branch of the family. It went always to the first-born. He decided then to call himself Joseph, and to resolve the confusion caused by being at the same time both Terence and Noel.

As a result of those early, unsettling domestic encounters he came to believe that an only child, without brothers and sisters in whom to confide, is provided, subconsciously, with an emotional basis for acting as a bridge-builder, 'searching always', as he came to put it, 'for a reconciliation, for a union of things separated'. He saw himself as 'always ferrying between two pieces of land separated by an arm of the sea'.

His mother, Alicia Adelaide, described herself as 'a daughter of music', the title she gave to her unpublished autobiography, which, though always at hand, her son never took the trouble to read. He knew her as a gifted but feckless person: a musician, pianist and composer with 'an artistic temperament'. She wrote that from the age of 5 she was 'a little musician, never tired of music or found any difficulty with it'. She was responsible for imprinting in 'Terence' a love of music. Her favourite composers were Chopin and Schumann; he came to prefer Beethoven and Mozart.

Alicia Montgomery was born in 1870 on Hallowe'en in Bangor, Ireland, where her father was town clerk for many years. Her mother came of a French family, the d'Argues, from the Arcachon region, near Bordeaux, who had settled in Co. Cavan. When, still a child, the family moved to Co. Down, where in the shadow of the Mountains of Mourne, under its highest peak, the great Slieve Donard, she indulged in her visions of the fairies, insisting always that 'Saturn has ever had his cold eye on me.' Nearby, in the old cathedral churchyard, was the authentic grave of St Patrick, the saint who had a lifelong influence on her.

She was the second wife of Dr Joseph Needham, whom she met on a visit to London in 1891. He had been a widower for eight years. Tragically, in one week he had lost his first wife, Robinetta Coghlan, daughter of the Vicar of Marchwood, near Southampton, and Josephine, his 15-year-old daughter. They had both died of diphtheria before the anti-toxin was available.

The doctor and the musician were speedily attracted to each other. After six weeks they became engaged and were married on 2 August 1892 in St Margaret's, the little church alongside Carlisle Cathedral. They spent their honeymoon in Scotland, where Alicia had family links. Her father was descended from the Montgomerys of Ayrshire, the family of the Earl of Eglinton, who had settled in Co. Down in the time of James I.

Dr Joseph Needham, in contrast to his flamboyant, extravagant wife, was a sober, scientific, sceptical person, with a great love of books. He was one of seven children (six of whom were boys),[1] in a poor family, by origin Huguenot weavers who had settled in Spitalfields in East London, where he was born in 1860, a Cockney, within the sound of Bow Bells. There were also links with Derbyshire, where the family name is quite common.

1. Interestingly, in the papers Joseph completed when he was elected a Fellow of the Royal Society he stated there were eleven children in all. But I have not been able to substantiate this.

In about 1880, he obtained his medical qualifications at Aberdeen University, which in those days provided financial help for students in needy circumstances. He earned a reputation as an anatomist, teaching at the university, and as a pioneer in pathological histology. He went into private practice, and by the time his son was born he was doing well financially as one of the first Harley Street specialists in anaesthesia.

Temperamentally, wife and husband were ill-matched. However, she wrote that 'the doctor was an artist in the love of everything beautiful and good and the house in London was a realm of good taste and luxury'. It was a tall, gaunt, but comfortable mansion at 34 Loats Road (today absorbed into King's Road), Clapham Park in South London. It was then an expensive, highly desirable residential area, where Mrs Needham wrote, 'Each spring there was a glorious show of chestnut, lilac, laburnum and hawthorn and one of the last windmills was near by.'

On the top floor was the nursery of Terence/Noel, where he was looked after by a Parisian governess. Although the room was filled with toys, building blocks, Meccano sets and an elaborate model railway, no other children, whether cousins or neighbours, ever set foot in it to play with him. An exception was Frankie Fahy, a neighbourhood child, who was allowed in on one occasion to have tea with him. Frankie was offered a cream pastry in the shape of a trumpet. He ate the cream, but not the pastry. He was never invited again.

Needham was a solitary, introspective, only child. Encounters with other members of his family were non-existent due, he insisted, to the snobbism of his parents. As they rose up the social scale, so did their insistence on strict Victorian standards of decorum. On one exceptional occasion he was taken over to Bangor to see his grandfather, but he remembered nothing of that. He did come to meet one of his father's brothers, walrus-moustached Uncle Alfred, who looked like Kaiser Wilhelm, a most genial businessman in a frock coat who had a successful furniture factory in East London. He was to help when Needham's father died in 1920 by taking care of financial matters until Needham was old enough to handle them himself. Uncle Alfred had a daughter, Etta, remembered as 'a really charming, affectionate and sensible girl, but in those days my mind was so closed to all ideas of girls, that I did not appreciate her as she deserved.' From time to time, he saw a step-uncle,

John Coghlan, a university scholar, who introduced him to the Greek and Latin classics.

His parents never took holidays together. Each claimed him for their own. He remembered his father teaching him to write with the usual 'pothooks' and 'hangers' as his mother hammered on the door insisting he was too young for that. She was then a successful song-writer, her work published by the leading music publisher, Novello & Co. Her 'Irish Melody' was a great hit around the world. She was also proud of the fact that during some St Patrick day festivities she had conducted the Band of the Irish Guards in 'St Patrick Was a Gentleman'. Her song, 'My Dark Rosaleen', was nearly adopted, instead of 'The Soldier Boy', as the national anthem of the Irish Free State.

Needham preferred the company of his father, who did not embarrass him as did his exuberant mother. Typical was the occasion 'Terence' was taken by her to the toy department of the fashionable Army and Navy Stores in London, where, expecting some inexpensive gift, he was overwhelmed to be bought a complete field ambulance unit, with tents and motor transport, nurses and surgeons. Then, the gift presented, he was as always left on his own to amuse himself. In later years, he wondered, 'Whatever happens to a boy's Oedipus complex when he grows up in a situation like that?' Needham could never provide an answer, but it lies in his later story.

As the only child, Needham was treated with special affection by his father, whom he came to see as a very old-fashioned bourgeois physician. He would speak of 'the lower orders', and of 'the genial bobbies', when he was police surgeon to W. Division of the Metropolitan Police. His father had a wealth of terse maxims which he used for teaching purposes and which Needham never forgot. Thus a labour-saving maxim was, 'Never go upstairs empty-handed, my boy'; a health one, 'Never have three helpings of anything, my boy'; against procrastination, 'Never put off till tomorrow what you can do today, my boy'; for orderliness, 'A place for everything, and everything in its place, my boy.' Needham would go over these, repeating them, as he pulled his wooden motorbus round the garden, stopping occasionally to cry out, *'Place des Invalides'* or *'Place du Louvre, tout le monde descend'*, which he had learned from his French governess.

His father encouraged him in developing scientific and mechanical interests. He would take him to Hampton Court with a collecting bottle to sample the stagnant waters, and to examine with a microscope blood vessels in a tadpole's tail or a frog's mesentery. He stimulated Needham's interest in models, especially railways, which he helped him to construct in elaborate layouts in the top-floor playroom. He gave him much training in the use of tools.

One of the wonderful aspects of his boyhood was the excellent library his father had collected. The books lined the walls of his consulting rooms and, inevitably, overflowed into the house. He was free to read whatever he wished, and his choice is a significant guide to the polymath he became. From about the age of 10, he remembered soaking himself in Schlegel's *History of Philosophy*, and acquiring a permanent love for Sir Thomas Browne's *Religio Medici*, and the other books he wrote in his 'marvellous seventeenth-century English', such as the *Pseudodoxia Epidemica, or Vulgar Errors* and the *Garden of Cyrus, or the Quincunx*.

He read also such books as *John Inglesant* by W. H. Shorthouse and *Preces Privatae* by Lancelot Andrewes. These were works with an Anglican ethos, to which he never lost his attachment. He was thrilled by Rawlinson's *Manners and Customs of the Ancient Egyptians*, which may well have fixed, even at that early age, a view that all the apparent absolutes of the traditions of Christendom were not absolutes at all, but formulations of relative value keyed to the particular forms of one civilization only. He loved the dignified dress and bodily beauty of Ancient Egypt, its solemn religion and its hieroglyphic script.

In religion, he did not get much from his mother, who was always rather vague on the subject, despite her love of St Patrick. His father, who was very conscious of the problems relating religion to philosophy and sociology, influenced him strongly. He had at one time been a leading figure in the Oxford Movement of Anglo-Catholicism.[2] Later, he was attracted by the Quakers, though he never became a Friend, and came finally to a kind of philosophical theology.

2.   A movement begun by J. H. (later Cardinal) Newman and John Keble, which tried to restore pre-Reformation religion to the Church of England.

It was this that led him to take the 10-year-old Needham to the Temple Church in central London each Sunday to hear its Master, the mathematician E. W. Barnes, a Fellow of the Royal Society, and later Bishop of Birmingham. For the young boy, all was excitement: to journey from South London by tram to the Embankment, to hear the bells of St Paul's Cathedral, and to walk to the Temple to listen to sermons which in fact were exciting lectures on pre-Socratic philosophers and medieval Scholastics. Typical subjects were 'Gnostics and Basilides', 'Mahayana Buddhists', 'Manicheans' and 'Eighteenth-century Deists'. Needham was exhilarated.

These and later influences had one extremely important effect on him: they 'liberated my mind from any connection between religion and that creepiness or spookiness which puts so many people off'. He came to be open to the personal experience of other religions, but above all to accept the importance of a rethinking of Christian doctrine and practice based on scientific knowledge, in regard, for instance, to sexual questions, race relations and social justice.

But such clear statements were still to come. Meanwhile, he grappled with them as he drove with his father in the family brougham to Dulwich College Preparatory School. His years there left little impression on him. What he did look forward to was his father in the carriage reading the *Fables* of La Fontaine in French, and explaining what that language was like in the seventeenth century. His father was an earnest francophile, who ensured that his son came to be as much at home in France as he was in England. He took him frequently to that country on holiday, and when Needham was 9 or 10 he spent a few months in a French school at St Valéry-sur-Somme. On one occasion, in the station restaurant in Boulogne, his father declared, 'You will never see chicken done like this in England.'

His father influenced him, also, by passing on punctual, regular behaviour. For instance, after school at 4 o'clock each day he would travel home on the white National Steam Bus, where, without fail, he would be seated eating a chocolate biscuit he bought always at the same shop, which on occasion he would share with the bus conductor. He preferred this unimaginative routine to what would have been his mother's frenetic behaviour in hunting out likely shops, where she would have demanded a display of chocolate biscuits and bought a dozen boxes immediately.

This ordered way of life was grossly interfered with in 1914 with the outbreak of the First World War. He was 'packed off', as he put it, to Oundle School in Northamptonshire. How the choice was made is not clear. He did not enjoy his four years there: he was very homesick, missing his intimate privacy in a public environment closeted with some hundreds of fellow boarders. However, in later years he came to recognize how fortunate he had been to have as headmaster the pioneering Frederick William Sanderson.

Oundle is one of the oldest of English schools. Its accepted history goes back to 1485, but the great change began in 1556 when Sir William Laxton, Sheriff and Lord Mayor of London and Master of the Grocers' Company, who was born at Oundle, bequeathed property he had in London to the company on condition they supported a school at his birthplace. The company agreed and continue to this day to provide the governors.

In September 1892, they took a bold step. They appointed F. W. Sanderson as headmaster. This was, in part, because the majority of governors were anxious to modernize the school's curriculum and to have scientific and technical subjects play a significant part in it. Sanderson had a sound reputation for his work in science and engineering as senior physics master at Dulwich College.

By the time Needham arrived at Oundle, Sanderson had provided the school with an educational ethos sufficient to cause H. G. Wells to write of the headmaster's 'vision of the school as a centre for the complete reorganization of civilized life'. He thought him 'beyond question the greatest man I have ever known with any degree of intimacy'.[3]

Sanderson came from a simple country background. He was not educated at a public school, but won a scholarship to Durham University as a theological student. He did well in that subject and in mathematics. From there he went to Cambridge, where he took a moderately successful degree in mathematics and worked for the Natural Science Tripos. He established a good reputation as a coach and was given a lectureship at Girton College, before moving on to teach at Dulwich College.

3.  H. G. Wells, *The Story of a Great Schoolmaster: Being a Plain Account of the Life and Ideas of Oundle*, London, Chatto & Windus, 1924.

Although education at Oundle was divided into classical and modern, it differed markedly from that of the traditional public school. This reflected Sanderson's educational vision. He was against boys working individually: he wanted them, as it were, to nourish each other by working in teams. He believed that as individuals they lacked the stimulus and help of a body of people working for the same end, the 'atmosphere was wrong'.[4]

This was why he divided the school into groups – classical, modern languages, science and engineering, and junior school. In this way, 'specialists' disappeared and became members of a form. His root idea was to stimulate co-operation. 'A failure at any point meant a breakdown in all. Each boy felt and knew that others depended on him; he was spurred to his best efforts, not by the spirit of competition but by the spirit of co-operation.'[5]

This scheme of co-operative working was extended to all fields of study. It was particularly applicable to applied science. In 1914, a new science block was built with ample space for large-scale experiments. It contained large chemistry and physics laboratories, a smaller biology laboratory, a drawing office and an impressive machinery hall, suitable for testing purposes.[6] The block was completed as war broke out. The hall with its workshops was able to make a serious contribution to the war effort in the manufacture of munitions. It was another example of Sanderson's insistence on the prime importance of constructive work for the community as opposed to exercise work for so-called instructional purposes.

The workshops came to play a prominent part in the lives of all at Oundle. Every boy, whatever his field of study, had to spend some hours each week in the metal shops. These contained a great range of machinery – lathes, milling machines, steam engines, a foundry for making metal castings. Woodwork was encouraged and tools of all kinds were made available. This enabled Needham to acquire a basic store of engineering knowledge, which was to serve him well.

There was, of course, the usual emphasis on sports and athletics, in which Needham did not take part. He had his father

4.   H. G. Wells, *Life*, London, Chatto & Windus, 1923.
5.   Ibid., p. 40.
6.   Ibid., p. 41.

'commit perjury' by testifying as a physician that sports activities were 'dangerous' for his son. There was a lot of bullying in Grafton, Needham's house, and a generalized, diffuse anti-intellectualism. That was why Needham disliked school work and found himself always in some unpopular minority.

Fortunately, he was able to take part in many of the activities of the Biology Sixth, and as 'a literary person' he found his way around: for instance, by becoming editor of the house magazine. He was helped by Sanderson's belief in the importance of library work. Boys were encouraged to read up any subject that interested them and then write an essay. Needham kept a long essay he had written on the rise and fall of the Knights Templars.[7]

There was, also, a very enlightened system for the study of the classics. This included not only due attention to Greek and Latin grammar and linguistics, but also encouragement to read such thinkers as Plato and Aristotle in English, so that as students they might arrive at a better understanding of their philosophy.

But the greatest influence on Needham during his four years at Oundle was that of the headmaster. He recalls Sanderson as 'a man of genius'. That was the view, also, of H. G. Wells, whose two sons were sent to Oundle on the outbreak of war. For Wells, he was an original, vigorous teacher, a pioneer feeling his way to 'a modernized education'. He described him as 'a ruddy, plethoric man, with his voice in his throat, and always very keen to talk'.

Wells was 'a great hero' for Needham. His parents always tried to get him to read the works of such classic writers as Jane Austen and the Brontës. But he preferred the science-fiction and social philosophy of Wells. He met Wells in person for the first time as a result of another of Sanderson's initiatives. A boy with a chemical or biological background or interest was encouraged to do voluntary work in those fields. One speech day, Needham was explaining details of nucleated and non-nucleated corpuscles of various mammals to a group of visitors, who included Wells. On that occasion, he met, also, Frederick Gowland

---

7. An order founded in 1138 by French knights to protect pilgrims to the Holy Land. It was suppressed in 1312 by the Pope for corruption and military insolence.

Hopkins, the distinguished biochemist, in whose laboratory in Cambridge he was to work. 'It was obvious', Needham recalled, 'that he knew quite a lot about blood corpuscles.'

Sanderson had a vision similar to that of Wells. He exhorted the boys to 'think in a spacious way, on a grand scale, laying aside the mean and the pettifogging'. To express this he had the idea of a House of Vision. This began as a memorial for Eric Yarrow, who was killed in the bloody fighting at Ypres in the First World War. He was the son of Sir Eric Yarrow, the great shipbuilder, who provided the funds for a museum of industrial history and organization. Architecturally, it turned out as a great disappointment for Sanderson, who was becoming more concerned with social and creative growth. Needham remembered the 'scripture classes' which were, in effect, discussions of Bible stories, primarily in terms of history and archaeology. Sanderson filled the building with objects from past civilizations, and as he believed in the importance of making historical charts, also with those constructed by the boys. These showed how cultures, civilizations and dynasties had come and gone and what their technological and intellectual achievements had been. Needham was responsible for a few of those charts. His original Oundle Bible has complicated diagrams of Assyrian, Babylonian, Hittite, Chaldean, Phoenician and Egyptian history pasted into its covers. He was much influenced by the significance of charts in story-telling. A dozen years later, one of his first publications was *A Chart to Illustrate the History of Biochemistry and Physiology.* Charts were to become of great value to him.

For Sanderson, a better knowledge of the past led to a better idea about the shape of the future. For Needham, it meant that a steady rise in social organization could be discerned as primitive tribal families united into city-states and nation-states, and these in turn increased in size with more and higher organizational structure. Such conceptions ultimately formed part of the basis of Needham's whole-world view, 'when I came to feel that social evolution was continuous with biological evolution and that, in turn, fitted in to cosmic evolution'.

He felt also that Sanderson's plea for 'spaciousness' coincided with ideas he had got from his mother. 'It rang a bell and I knew instinctively that it was right.' Without that, he insisted, the largest work that he was to do would never have been conceived or attempted. 'My father would have been too cautious, but Sanderson struck the same note as my mother and I resonated strongly with it.'

Although the depth of Sanderson's influence on Needham was great, he had very little personal contact with him, apart from listening to him as a pupil in class. The only face-to-face interview with Sanderson he can remember was when he was leaving to go to King's College, University of London. Needham found it 'a bizarre circumstance' that his only memory of what Sanderson said was, 'Well, you'll never do anything to disgrace the school, my boy.' This comment puzzled him, because he felt that Sanderson should not have left him with such a very negative impression.

Whilst at Oundle, he had trouble with his teeth. The nearest competent dentist was in Peterborough, about thirty miles away. As a result he had to spend many hours waiting for the train at the now long-gone Peterborough East station. There he showed so keen an interest in the shunting and movement of the trains that Alfred Blincoe, 'a kindly old man, driver of the shunting tank-engine there, befriended me, took me up into the cab, and he and his fireman gradually taught me the principles and practice of driving a steam-locomotive, so that eventually I could (quite illegally) take over the regulator and the Westinghouse brake, and crack a walnut (as they say) gently between the buffers'.

He learned two lessons from that experience. First, it exemplified a dictum of his father's that no knowledge should ever be wasted or despised. Second, it brought him 'into union and sympathy with working-class people', something that his parents' bourgeois class-consciousness might otherwise have prevented. It reinforced an experience he had had when he was about 13. On holiday with his father in France, they had missed making a train connection at a little railway junction called Eu. They had to stay the night and the local hotel was full. His father was worried, but a railwayman 'invited us to his humble home and made us most welcome there'.

Through his father, who was medical officer of a detachment of the Surrey Yeomanry, a volunteer cavalry regiment, he learned as a child to ride a horse at the regimental riding school opposite his home. At about the same time, aged 8, he began to use the family Yost typewriter. He became so proficient at this, finding it easier to compose directly on to the machine than to use longhand, that throughout his working life he would type his own manuscripts. He would also do his own filing.

While Needham was in his final year at Oundle, the war and its slaughter went on. His father was busy from morning to night as anaesthetist at three large London military hospitals: the King George, near Waterloo Station, mainly for brain surgery; at the Third London General in Wandsworth for special plastic surgery; at the Weir Hospital in Clapham for general cases. So great was the need for operating theatre assistants that Needham became involved during his school vacations, from the age of 16, working from dawn to dusk in operating theatres – handling instruments and catgut like an assistant theatre sister. It was very different from the first surgical operation he had attended when he was 9 years old – an appendectomy carried out by Sir John Bland-Sutton, who gave him a gold sovereign for assisting.

This experience produced in Needham a feeling of particular usefulness. Although he developed a splendid knowledge of anatomy, and it was 'marvellous to be able to see the results of successful surgical work', he came to realize that surgery held no intellectual interest for him.

But the demands of war caught up with him. In 1918, the lack of qualified medical personnel in the services was so alarming that even medical students who had not taken any examinations were rushed through crash courses of first aid and gazetted as second lieutenants in the Royal Army Medical Corps, or, as in Needham's case, surgeon sub-lieutenant, Royal Navy Volunteer Reserve, though he was never required to serve at sea, as the war ended in November.

Before that, he had a few months at King's College, London. He performed his first anatomical dissection there, 'down in the depths' as he put it; and then 'high up under the roof' he acted as laboratory assistant to the great biochemist, Otto Rosenheim. He learned how to prepare quantities of lecithin, kephalin and sphingomylin, the lipids of the brain, on which Rosenheim was a world authority. Needham commented, 'It was a good introduction to laboratory research. But study at Cambridge awaited me.'

# 2

# Egg and embryo

*Laboratorium est oratiorum*
*(The place where we do our scientific work is a place of prayer.)*

Joseph Needham

On 11 November 1918, Needham drank his War Victory Toast at Cambridge as an undergraduate at Gonville and Caius, a college with which he was to be most intimately bound, in a city and region which were to envelop him. There were not many present in Hall on that occasion, but he sensed a gulf between those who came up fresh from school and those who had served in the armed forces. He did not feel himself to be in either group, mainly because of his highly personal hospital experiences.

He was in the only Oxbridge college with a double name. Its founder in 1348 was Edmund Gonville, a well-connected, Norfolk parish priest, prospering in managing land and tithes. John Caius (pronounced Keys), the other founder, appeared almost 200 years later, in 1529, aged 18, as a scholar at Gonville Hall. He obtained his B.A. early in 1533 and was elected a Fellow later in that year. He spent a further two years completing his M.A.

He became a distinguished physician, with sound connections at the court of Elizabeth I. In 1558, he received the Queen's approval to re-found and re-endow Gonville Hall. A Royal Charter converted it into the College of Gonville and Caius, founded in honour of the Annunciation of the Blessed Virgin Mary.[1]

Over the centuries, following the example of the founders, the college produced many medical men and theologians. The

---

1. Christopher Brooke, *A History of Gonville and Caius College*, Woodbridge (United Kingdom), Boydell Press, 1985.

greatest was the physiologist, William Harvey (1588–1657), the discoverer of the circulation of the blood, and physician to Charles I.

Needham entered Caius, as the college is familiarly referred to, by chance. His father had done no preliminary research so had no special recommendation. However, while lying in the long grass near Oundle during Officers' Training Corps open-order drill, he found himself next to Charles Brook, who, knowing that he was due to study medicine at Cambridge, asked him what college he was going to. Needham admitted he had not the faintest idea. Brook suggested he try Caius, a college with a good medical reputation, which was where he was going. Needham applied and was accepted. The Master in 1918 was Sir Hugh Anderson, an eminent neuro-physiologist, whom Needham found kind and encouraging. Those qualities were highly appreciated by him in his depressing room, No. C1, on the ground floor, in extremely gloomy St Michael's Court.

It was there that he went down quite badly with influenza in the devastating post-war epidemic. The only tutor in residence, described as 'the College Pooh-Bah', was acting also as dean, praelector, and steward. He was V. T. Vesey, an Irishman, who had changed his name from Lendrum when he inherited a country estate. He was an accepted Cambridge eccentric, who once every week in the hunting season was to be seen marching out of college in scarlet hunting costume.[2]

Needham recalled that Vesey brought him a bunch of grapes which, to avoid infection, he offered attached to the end of a walking stick. What he enjoyed when he had recovered was the breakfast from the college kitchens, brought each morning to each student separately on a big wooden tray covered with a green baize cloth. He always had a sweet tooth, and he also cherished the delicacy, *petit jambon,* made of a little tower of fried bread containing creamed ham topped by a poached egg. In the rooms of his friend William Brockbank, who was to follow his family medical tradition and become a consultant physician at the Manchester Royal Infirmary, he shared the delicious north-country spicy, ginger cake called 'parkin'. Needham's other great friend Monty Maybury was, like him, an Anglo-Catholic medical student. He became a general practitioner in Portsmouth and

2.   Brooke, op. cit., p. 237.

raised a large family of boys who all became doctors. As far as sports and athletics were concerned, as at Oundle, he showed little interest, but joined the 'Caius Unemployed', the least of the college rugby teams. They were called 'unemployed' as a mark of solidarity with the many unemployed in the country as a whole.

The immediate post-war years were of unique importance in opening up the affinities between the physical and biological sciences. Biological phenomena were being revealed as performing in accordance with physical laws, including those of chemistry. Most fortunately, Needham's director of studies, the black-bearded, romantic-looking Sir William Bate Hardy, listened to the shy, widely read, religion-bound young man, who in his solitary explorations from Cambridge was already walking his way along the side roads of the past to build a number of bridges linking trodden highways and virgin pastures.

Hardy cautioned him against pursuing only biological studies, such as anatomy, physiology, and zoology, in preparation for a medical career. 'No, my boy,' he would state firmly, 'that won't do at all. The future lies with atoms and molecules, my boy, atoms and molecules. You'll never do anything in biology if you don't have that chemical and physical basis. That's what you should study.' Needham was, as he said, 'pitchforked into chemistry, anatomy and physiology', which affected his whole future development.

He had little recollection of those introductory days, and remembered clearly only one of his supervisors, Sir Rudolph Peters, his director of studies and supervisor, who left Caius in 1923 to become Professor of Biochemistry at Oxford. He was to stimulate Needham a decade later in some highly original work. On his retirement from Oxford, he returned to Caius as an Honorary Fellow.

Needham obtained his B.A. in 1921, and in that year went to Freiburg im Breisgau in Germany, to do research under Professor Knoop. But in those difficult, immediate post-war years he found laboratory work there rather disorganized. Also, his German was poor. Accordingly, he spent most of his stay in absorbing, in his typically thorough manner, the German language, in which he became fluent. He also made an approach to an understanding of the wider German culture. To his great delight, expecting post-war, anti-British feeling, he found the Germans he met extremely kind and helpful.

In 1922, Needham was admitted as a postgraduate student to the Cambridge Biochemical Laboratory, then directed by the much-loved, highly regarded Frederick Gowland Hopkins, who, despite his knighthood in 1925, was always popularly known as 'Hoppy'. Biochemistry was then at an interesting stage of development. It had begun to reshape itself towards the end of the nineteenth century by transformation from physiological chemistry. For some four decades, from about 1880 to about 1920, the development of enzyme chemistry had provided a basis for a dynamic biochemistry. When Needham began his research work, he was able to contribute to the great increase in studies of structural and dynamic relations in biochemical changes. His major speciality came to be the biochemistry of embryonic development, that is, research into sequential evolution in chemical processes, how, starting from a single, fertilized egg cell, the developing foetus acquires its highly specialized parts – heart, limbs, eyes, etc.[3]

It was 'Hoppy' who finally caused him to give up all thought of a medical career and to do research on the borderline between biochemistry and experimental morphology (the science of the forms and structure of organized beings) and embryology. The ultimate aim of this, as J. B. S. Haldane was to put it, was to present a 'complete account of intermediary metabolism, that is to say, of the transformations undergone by matter in passing through organisms'.[4]

The intellectual challenge that Needham had found absent in surgery, he now discovered with great excitement in 'the enormous problem of seeing the connections between the chemical level and the morphological form, the atoms and molecules on

3.  William Harvey had strong views about the embryonic origins of animals. He wrote in 1651, 'all animals, even those that produce their young alive, including man himself, are evolved out of the egg'. By 'evolved' he meant 'developed'. He got permission from deer-hunter King Charles to dissect the shot females and to look for embryos. He was able to find earlier and earlier stages of the developing embryo in the uterus, but not the earliest stage. Having no microscope to see an egg, he may have suspected it was very small.

4.  J. B. S. Haldane, *Perspectives in Biochemistry*, London, Cambridge University Press, 1937.

the one hand and the organs and tissues of the body on the other, indeed the whole structure of the developed human organism'.[5] That raised agitating philosophical questions, and problems that did not arise for those who worked solely on the chemical or anatomical levels.

From 1920 to 1942, Needham's 'home' was the Biochemical Lab – as a student, a researcher, a demonstrator, and lastly as Sir William Dunn Reader. The father-figure for all Cambridge biochemists of Needham's own and neighbouring generations was Hopkins, the true founder of modern biochemistry in England.[6] 'All of Hoppy's geese', went the saying, 'turned into swans.' 'Hoppy' ruled his laboratory with love and tenderness. He never allowed his name to appear on a publication unless he had been involved in some of the practical work. Unlike many heads of department, he would not give out problems to researchers. It was understood that if a student knew what he wanted to do, 'Hoppy' would back him fully. But if he did not know, he would let him float away.

Needham began his research at a time when chemistry was becoming an effective way of approaching biological problems: when study of the cell, through the development of new instruments, such as the phase and interferometer microscope and the ultra-violet and infra-red reflecting microscope, revealed it as a highly complicated but ordered structure.

His early work was on the metabolism of inositol (a crystalline substance found in muscle and urine) and other cycloses (a class of carbohydrate which includes inositol). The function of inositol in the body had long been a mystery. His papers on this, published from 1923 onwards, were, as he put it, a *Jugendarbeit* (young person's work), though some interesting facts came to light. For example, he was able to confirm the isolated finding by a young German, named Klein, who probably died in the First World War, who stated that a hen's egg at the beginning of development contained no inositol, but by the time it hatched

5. 'Interview with Joseph Needham', *The Caian*, 1976, p. 40.
6. 'For nearly twenty years now truly in loco parentis to me,' wrote Needham in 1936 in his book *Order and Life*, p. 140. The story of Hopkins, the creator of British biochemistry, is well summarized in J. G. Crowther, *British Scientists of the Twentieth Century*, pp. 197–247, London, Routledge & Kegan Paul, 1952.

several hundred milligrams had been synthesized. Needham thought what a marvellous chemical factory the developing hen's egg must be during its three weeks of incubation.

He consulted 'Hoppy', who gave him full support to investigate. 'Hoppy' told Needham that when he came to Cambridge as a biochemist in 1896, he was put up overnight by Sir Michael Foster, the professor of physiology. At breakfast, Sir Michael said, 'Now, Hopkins, what you ought to do is to discover how the marvellous red pigment, the haemoglobin, is made during the development of the hen's egg. There is none there at the beginning, and yet there is plenty of blood at the end.' 'Hoppy' himself never did exactly that, but he may have felt that Needham's work might help to elucidate Sir Michael's comment. Needham never forgot how he came to understand 'Hoppy's' fascination with the hen's egg as a closed system of chemical transformations.

In collaboration with Dorothy Mary Moyle, a fellow biochemist, whom he came to marry, there began a series of studies on the cell interior: on the way in which, for example, organisms as they evolve rid themselves of their nitrogen products in the form of ammonia, urea and uric acid. Dorothy was one of a group of brilliant women research workers who, through the encouragement of 'Hoppy', entered the Cambridge Biochemical Lab after 1918. She was attached to the Sir William Dunn Institute of Biochemistry from 1920 until 1963.

They found they could combine business with pleasure by working through the vacation weeks at different marine biological laboratories. At Millport Marine Biological Station at Keppel Pier on Great Cumbrae Island in the Firth of Clyde, they elucidated the nitrogen metabolism of developing dogfish eggs, and they worked a number of times in the United States at the Woods Hole Institute in Massachusetts, and on the phosphorous metabolism of invertebrates during embryonic life at Monterey in California. At Roscoff in Brittany, they performed micro-injections of the eggs of invertebrates before and after fertilization.

They studied the pH and oxidation-reduction potential of the cell interior by means of micro-manipulative technique. This necessitated the mastering of a technique so difficult that the Needhams had few competitors. However, it did not lead to as great an insight into cell physiology as was at first hoped.

There followed a long series of projects on the metabolism of the developing egg, mainly the chick, but also of all kinds of invertebrates. Some of them were published under the title, *Energy Sources in Ontogenesis.* Needham hit upon the interesting rule that there is a succession of energy sources during development: carbohydrate preceding protein, and protein preceding fat. The only uncertainty then was whether this started from fertilization or from gastrulation (the process in which the embryo is transformed from the blastula stage by invagination, that is, infolding of the blastula wall, to form the key cap-shaped gastrula structure). As he wrote, 'Judging from the number of workers in America and elsewhere who have since confirmed this rule on the most diverse variety of embryos, both of vertebrates and invertebrates, it must be one of very wide, if not universal, validity. I cannot but regard it as one of the most far-reaching generalizations which has arisen from my own experimental work.'[7]

In addition, he was able to describe a large number of new facts about the metabolism of embryos and the chemical functions of their accessory structures during development, such as the yolk-sac and allantois (umbilical blood vessels). He found the former acting like the mammalian placenta as a transitory liver. One series of papers was devoted to the strangely large differences in osmotic pressure, pH, etc., between the yolk and the white in a hen's egg.

In 1931, Needham's seminal three-volume work, *Chemical Embryology*, was published. It was entirely original in that it had no predecessor save a brief work, *Spezielle Physiologie des Embryos* [Physiology of the Embryo] by W. Preyer, published in 1885, which had nothing to say about biochemistry. His anxiety to paint a logically complete canvas led him, typically, to press on irrespective of bulk, though he had planned a more modest work. As it turned out, practically nothing was overlooked: it consisted over 2,000 pages, with a total of almost a million words. His theme came to him from Samuel Taylor Coleridge, the eighteenth-century poet, critic and philosopher, who wrote, 'The history of a man for the first nine months preceding his birth would

---

7.  Memo, personal records of Fellows of the Royal Society, January 1941.

probably be more interesting and contain events of far greater moment, than all the three-score years and ten that follow it.'

It was not only in line with his fascination for the history of science that he felt it necessary for the first volume to carry a substantial 200-page survey of the history of embryology, in which he presented a remarkable collection of views, quoting original sources in English, German, French, Italian, Greek, Latin, Spanish and Arabic, on the development of the human foetus throughout history. He dealt also with the involved question and desperate theological and social implications of the implantation of a soul in the foetus. It was clear that even at the age of 31 he was well qualified as a polymath. He argued, also, that such a book was more valuable for subsequent workers than almost any experiments one could do: for instance, few remembered the detailed experiments of Wilhelm Roux, regarded as the first real experimental embryologist. He interfered with the development process: for example, his 1888 experiment to test the idea that the 'potentiality' for an entire frog embryo must be present in the organization of materials in the egg. When the fertilized egg divided into two cells, he used a hot needle to kill one of the cells. He found the remaining cell stayed healthy and continued to divide.

Needham's point was that although Roux's experiments were important, through writing the book he first grasped the complete possibilities of the experimental method in embryology, and in a similar, but minor, way he did have a glimpse of the range of biochemical shifts involved in embryonic development: the matter and the form both changing with time.

Occasionally, he felt that the book was sometimes regarded (by those who had not read it, he would stress) as a mere compilation or collection of literature. On the contrary, he insisted, during the writing of it he was led to make many correlations, which previously had gone unnoticed because no one had looked at the field as a whole.

This was the view of Julian Huxley, who in a review wrote:

This is in its way a classical book. It demonstrates the extent of ground won by the pioneers in this new field, defines its scope, consolidates one part with another, and proclaims to the world the title of this young branch of biology from now on to sovereign rights in a territory of its own. . . . It has vindicated the claims of chemical embryology to independent and fruitful existence and marks another milestone on the road that biology is taking in its transformation from a statistic to a dynamic science. Biologists are under a very real debt to Dr Needham for his wide reading, his critical and constructive facility, and his patient industry which has issued in this fine work.[8]

Interestingly, Needham, who still kept in touch with his mother, had sent her the three volumes. She replied most emotionally in a telegram and a letter, both dated 1 December 1931, expressing 'many tears of gladness, unabashed pride, gratefulness'.

Of matters previously overlooked, he felt perhaps the widest was the conception of the cleidoic egg, such as that of birds and some reptiles, insects and gastropods, the walls of which were permeable only to matter in a gaseous state. That led to the only explanation of the origin of uricotelic metabolism (having uric acid as the chief nitrogenous excretory product) that had ever been given, namely, that it was necessitated by terrestrial oviparous life.

Another problem that required quite a struggle for its classification was the vexed question of *Entwicklungsarbeit* (development work), which before he wrote his chapter on it was in complete confusion. The question was how cell differentiation in the adult animal arose from cells of the early embryo which appear so much alike.

The approach to an answer began in spring 1924, when the German researcher, Hans Spemann, obtained the first induction, which led him and his young Ph.D. student, Hilde Mangold, to a fundamental discovery in 1931. It turned out to be a breakthrough in the borderline field of biochemistry and experimental embryology. It gave the promise of the possible identification of the molecules which act as morphogenetic hormones, or 'fate fixers', as they provided a driving force to certain groups of cells in the developing embryo to cause them to follow through a particular line of morphological and histological development. What was the 'Spemann organizer', that is, the part of the embryo that so influences some other part as to bring about the development?

This led Needham to a new line of work during the next ten years in collaboration with C. H. Waddington and others. It resulted in the publication in 1942 of *Biochemistry and Morphogenesis*, a general survey of the field. It was a seminal work, afterwards reprinted.

8.  *Nature,* 6 February 1932.

In 1930, Needham had gone to Brussels to learn the technique of embryological micro-operations in the laboratory of Albert Brachet, and in 1933, by linking up with Waddington, he was able to push ahead twice as fast as he could alone. Together with Dorothy, they worked for a period at the Institute of Otto Mangold in Berlin, each taking a hand in all aspects of the required technical work. From a theoretical point of view, the research on the biochemistry of organizer phenomena was of significance for regions of biology even outside embryology, wherever stimulus-response reactions were involved.

The work of Needham, Waddington, Spemann and others revealed that a non-living substance can act as an organizer for the embryonic differentiation of a nervous system. In *Biochemistry and Morphogenesis* (p. 172), Needham stated that a piece of boiled mouse heart placed among the living cells of a human embryo induced the formation of a secondary or 'extra' brain. This was of philosophical significance for Needham, for it revealed that there was a physico-chemical explanation for what had been regarded as the work of an all-knowing deity.

This affected him directly, for there were specific 'religious' activities in which he was involved. He was a member of the Sanctae Trinitatis Confraternitas, a society that organized plainsong liturgies in various college chapels and which heard papers on Church history. He became, also, the secretary of a university society, the Cambridge branch of the Guild of St Luke, an Anglican confraternity for doctors and medical students, which held monthly meetings. Needham was responsible for inviting 'great scholars', from whose addresses he said he learned more than from all the regular lectures he attended. He was inspired, for instance, by F. S. Burkitt on 'Manichaeism' and by Edward Browne on 'Medicine in Persian and Arabic countries'.

It was such men who 'first gave me an idea of the grandeur and excitement of humanistic scholarship'. This led him directly to the realization that while the natural sciences themselves were a wonderful activity of the human mind, their history and how they had developed through the ages from small beginnings was at least equally worthy of study.[9]

9.  Lu Gwei-Djen, *Explorations in the History of Science in China*, Shanghai, 1982.

Whilst doing research at the Biochemical Lab, Needham was for two years a lay brother at the Oratory of the Good Shepherd, whose headquarters were then in Cambridge. The oratory was an Anglican religious order modelled on that founded by the Cardinal de Bérulle, but with a special devotion to Nicholas Ferrar, the great seventeenth-century Anglican from Little Gidding in Huntingdonshire. Needham had hopes that other young scientists would become lay brothers and form a permanent group, but he was disappointed when that never happened. In this field, Needham was a pioneer in testifying to the need to unify the experience of science with that of religion.[10]

During the period Needham lived at the Oratory House, he got to know the Oratorian Fathers very well and became fond of them all. John How, the Superior, who became Bishop of Glasgow, taught him 'not to be blown about by every wind of doctrine'; Eric Milner-White, Dean of King's College, was an aesthetic liturgiologist with exquisite taste which was reflected in his understanding of music, literature and art; Wilfred Knox, who could combine the widest-ranging, freest intellectual speculation with regular mainstream Catholic practice and sensibility; and Edward Wynn, then Dean of Pembroke College and later Bishop of Ely.

He learned a great deal from the Oratorians, many things which he did not find easy to put into words, but which could be expressed in epigrams: for example, that 'Nothing is ever merely anything', as Samuel Butler once remarked. But he was finding that celibacy was not for him, and in 1924 he left the Oratory.

In September of that year, he married his colleague, Dorothy Moyle. As the wedding took place just before he was elected a Fellow of Caius, he recalled, the newly-weds missed having a wedding present from the Fellows, which, strangely, they felt rather badly about. Dorothy was known to her intimates as

10. The American cosmologist, Frank J. Tipler, in his latest book, *The Physics of Immortality* (New York, Doubleday, 1994), writes, 'It is unique to find a book asserting . . . that theology is a branch of physics, that physicists can infer by calculation the existence of God and the likelihood of the resurrection of the dead to eternal life in exactly the same way as physicists calculate the properties of the electron.'

Dophi. She was four years older than Needham, born in London on 22 September 1896. Her family was Cornish from the neighbourhood of St Austell, her father, John Moyle, being a minor civil servant working in the Patent Office in Streatham, London. He was a socialist, as was Dorothy. She was educated at Claremont College, Stockport, and Girton College, Cambridge.

Dorothy became an acknowledged international authority on the biochemical processes occurring in muscular contraction, on carbohydrate metabolism and the transfer of energy by means of phosphorylation processes. She began her researches under the inspiration of 'Hoppy', who said:

If one could get to the bottom of how cells used their energy, how they brought their energy to bear upon the job of contracting the muscle, which you could measure in mechanical terms, if that once could be found out, then the method of energy provision in all cellular activity, whether it was secreting or flashing lights, or doing any of the things that cells do, like keeping up osmotic pressure or something like that against the environment, would all be found to have the same mechanism.[11]

That came true, but Hopkins did not live to see it. Dorothy told the story in her book, *Machina Carnis: A Century and a Half of Muscle Biochemistry*, published in 1971. In 1948, she became one of the earliest women Fellows of the Royal Society, and with Needham the Society's first practising scientists, husband-and-wife team (the only other couple to have received the distinction was Queen Victoria and Prince Albert, obviously a special case).

Both Dorothy and Joseph had common political and religious interests as Christians and as socialists. Needham appreciated her highly as 'not being a flamboyant character, but a reasonable person, never *outrée* or unusual'. As Jennifer Williams, who met Dorothy in the early 1950s as a first year undergraduate studying biochemistry, put it, 'It was a privilege to meet her and to be taught by her. She had a very gentle manner, which belied her strength of mind.'[12] She was then recovering from

11. Text from a BBC interview with Peter Cantor, in December 1966, but not used in the broadcast.
12. *The Independent*, 11 January 1988.

tuberculosis, contracted while working in China with Needham during the Second World War. Jennifer, after graduating, worked as her assistant in the Biochemistry Lab on the contractile proteins from smooth muscle. Dophi, she wrote, 'worked very hard and for long hours. She was endlessly patient.' She drove around at a very sedate pace in a small, grey electric car.

There were no children of the marriage, and none from the extra-marital relationships they commonly agreed to allow. This caused them sadness, and a recognition of the fact that with Needham's death his contribution to the family would end. Each had many affairs, but that was not then uncommon among many married academics. Needham was always exhilarated by the presence of women, and he was fortunate because the Biochemical Lab was exceptional in those days in that female research workers were accepted on equal terms with males. Thus, he worked side by side with such personalities as Margery Stephenson, founder of bacteriological chemistry, and Dorothy Jordan-Lloyd, a brilliant protein chemist.

Throughout the period of his experimental work, he had from time to time written papers that were half reviews, half discussions of theoretical points not noticed previously. These had appeared usually in *Biological Reviews*. Thus, about 1929 he urged that 'recapitulation' (repetition in the individual in its development and growth of the evolutionary stages through which the species evolved) could not explain anything, but could itself be explained if the recapitulated structures gave morphogenetic stimuli of an organizer character for structures arising later. That prediction was verified by many workers for organs such as the chick pronephros (the primordial kidney).

As the Second World War entered its third year and important moves were to develop for Needham, he wrote that the publication of *Biochemistry and Morphogenesis* had enabled him to approach more closely than before a fundamental aim, essentially his, namely, the rapprochement of biochemical and morphological science. That would 'necessitate the replacement of the classical concepts of form and matter by those of organization (at many different levels) and energy'.

The work of two Caius Fellows, Sir William Bate Hardy, formerly secretary of the Royal Society, and Sir Rudolph Peters, influenced Needham greatly. Hardy began his own research career with the purest of zoology and morphology and ended with the physics of lubrication and refrigeration. This arose from his conviction that the problem of biological organization would never be solved by mere description of form at the higher levels, but by a bridge between the largest organic molecules and the smallest intracellular structures. That was what led him to become a physicist.

Peters proposed the concept of 'co-ordinational biochemistry': that is, a cell surface may be made of molecules so anchored as to constitute a chemical mosaic. His concept was what Needham meant by the extension of morphology into biochemistry and the bridging of the gulf between the so-called sciences of matter and of form. As a result, a logical analysis of the concept of organism led researchers to look for organizing relations at all the levels, coarse and fine, of the living structure. Biochemistry and morphology, Needham forecast, would then blend into each other.

However, something that had happened many years before now bore fruit. At the end of his third year of research in biochemistry, he had asked 'Hoppy' whether he could go in for the Benn Levy Studentship in biochemistry. He was told he must wait until graduation. He did get the studentship of £250 a year, a splendid sum in those days, and became Demonstrator and finally Reader in the Department of Biochemistry. He was fortunate that the award finally made it clear he would not need a medical degree to earn a living. What could then not be made clear was that by staying on at the Lab he would meet in 1937 three Chinese graduate scientists who would change his life.

# 3

# Integrative generalist

*Can any man be a good Naturalist,*
*that is not seene in the Metaphysicke?*
*Or a good Moralist, who is not a Naturalist?*
*Or a Logician, who is ignorant of reall Sciences?*
*Or a Divine, a Lawyer, or a Physician,*
*that is no Philosopher?*
*Or an Oratour or Poet,*
*who is not accomplished with them all?*

Comenius, *A Reformation of Schooles*, 1642

Needham was often accused of being like a magpie, endlessly gathering information. Certainly, he had an extraordinary facility for rigorous control of notes, usages and references, for the discipline of filing, and the maintenance of card indexes, all these overseen by his excellent memory, long before the word processor and computer came to easy hand.[1]

He had a special organizational gift, without which none of his creativity could have expressed itself. This is demonstrated in his books and many essays which he described as 'exciting reconnaissances, never saying the last word on anything, but opening up mines of treasures which other scholars can develop later, amending, correcting and expanding'. He was concerned always with presenting a systematic position in the practice of intellectual bridge-building.

He suffered for this daring. Traditionally, a professional scientist was required to stick to his speciality, and not to become regarded as a generalist. The scientific worker was obliged to remain alone in his mental prison, labelled – as he still is – by his area of speciality, and if he found time for extra-laboratory activity it should preferably take the form of a sport, such as golf or fishing or bridge. As such, argued the Scientific Establish-

---

1.  The story goes that Dophi was telling a colleague about Needham's photo-memory. When proof copies of a volume were at hand, she explained, Needham had a new game: 'He used to lie in bed correcting the proofs in his head. But he got bored with that. So now he translates them into French first – in his head, of course!'

---

ment, the stability of society would not be threatened, as it was by the young, socially responsible radicals of the 1920s and 1930s, such as Needham himself, J. D. Bernal, J. B. S. Haldane, L. Hogben and Hyman Levy. And so was born Henry Holorenshaw, Needham's *alter ego,* a ploy to ensure that the Royal Society would not deny him the important recognition bestowed by a Fellowship.

At an early age, Needham began to formulate a view of development which enabled him to see the future as a step-by-step evolution to an 'immortality' in which, for instance, there was no contradiction between being devoutly religious in a special sense and rigorously scientific. He felt that the world in which we live and work did not come ready made, that it was up to each person to reconstruct it.

His early *philosophia perennis* was based on the views of R. C. Collingwood, the Oxford scholar he never met, author of *Speculum Mentis* or *The Map of Knowledge* (Oxford, 1924), whose *Autobiography* (Oxford, 1939) revealed a development parallel to Needham's. He was much influenced, also, by Rudolf Otto, the German theologian, whose book *Das Heilige* (the sense of the holy one) was concerned with delineating the numinous, the meaning of religious wonder.

Collingwood's *Map* was 'the outcome of a conviction that the only philosophy that can be of real use to anybody at the present time is a critical review of the chief forms of human experience'. He distinguished five forms: religion, science, history, philosophy and aesthetics. For Collingwood, all thought existed for the sake of action. And as Needham commented, 'If you are tone deaf to one of the forms, you are in trouble.'

When he was younger, he felt that it did not matter if the five forms of human experience 'contradicted each other flat'. In those days, he was determined to be a 'divider' and almost welcomed characteristics that added to their irreconcilability.

His research in the 1920s and 1930s was done in a period when, to quote A. N. Whitehead, whose philosophy, expounded in his 1941 *Science and the Modern World*, influenced him greatly, 'Science is taking on a new aspect, which is neither purely physical nor purely biological.' He came to the view that science, as with all cultural enterprise, is in part a social construction. He recognized, also, 'the fact that the scientific investigator works 50 per cent of his time by non-rational means . . . often the most successful investigators of nature are quite unable to give an account of their reasons for doing such and such an experiment, or for placing side by side two apparently unrelated facts. . . . Experiments confirm each other, and a false step is usually soon discovered.'[2]

Needham believed that 'Francis Bacon, the official herald and announcer of the scientific method, totally misunderstood this quality . . . that the scientific worker operates to a high degree unconsciously. He imagined that the scientific method was itself as mechanical a thing as the picture of the universe which it produces by its efforts.'[3]

For Needham, a key road to understanding lay in the history of science, which he treated always with great passion, though he never had any formal teaching in that discipline. This enthusiasm originated in part from the influence of Sanderson, his Oundle headmaster, who spoke always with great enthusiasm of 'the spacious sweep of history'. As a student, Needham was enthralled by such books as Michael Foster's *History of Physics*, and Sir William Dampier-Whetham's *History of Science*.

Without history, he argued, the scientist would know nothing of social evolution, of the origin and progress of human society, of the laws of change and of the direction in which further progress was likely to take place. Without philosophy, the scientist could have no basic world-view, and might fall into all kinds of fantasies – for successful scientific work was compatible with anything from Roman Catholicism, as in the case of Pasteur, to Sandemanianism, a religious sect developed in the eighteenth century by Robert Sandeman, as in the case of Faraday.

2.   A. N. Whitehead, *The Sceptical Biologist*, pp. 80–1. See Lewis Wolpert, Professor of Biology as applied to Medicine at University College, London, in *The Unnatural Nature of Science*, London, Faber & Faber, 1992: 'It is often held that science and commonsense are closely linked. . . . However reasonable such views sound, they are, alas, quite misleading. In fact, both the ideas that science generates and the way in which science is carried out are entirely counter-intuitive and against commonsense – by which I mean that scientific ideas cannot be acquired by simple inspection of phenomena and they are very often outside everyday experience. Science does not fit our natural expectations.'

3.   Ibid., p. 2.

It was in the mid-1930s that he with some others succeeded in laying the foundations for the history of science and medicine as an independent discipline in teaching and research at Cambridge. This began in 1936 when the botanist, Hamshaw Thomas, organized in the Old Schools a loan exhibition of historic scientific apparatus collected from the Cambridge scientific departments and colleges. In the summer of that year, the Faculty Board of Biology finally agreed, following Needham's continuous nagging, to appoint him as a committee of one to co-opt others and to organize a History of Science Lecture Scheme. His first committee included Hamshaw Thomas; Dampier-Whetham; the distinguished crystallographer and Marxist, J. D. Bernal; the historian, later Master of Peterhouse, Herbert Butterfield; and the economic historian, Michael Postan. The secretary was Walter Pagel, then a tuberculosis pathologist, but later famous for his Helmontian and Paracelsian studies. The first course of lectures was very successful and appeared as a book, *Background to Modern Science*. From these activities arose the History of Science Department at Cambridge, associated with the local Whipple Museum of the History of Science.

His 1931 book, *Chemical Embryology*, was based on lectures he had given in the previous year at University College, London. They had been entitled, 'Speculation, Observation and Experiment as Illustrated in the History of Embryology'. In those lectures he had insisted on developing the societal relations between biological and embryological research.

He was conscious of the necessity to combine the history of science with economic and social history. He wrote:

That rather sharp cleavage between the philosophical biologist of the Hellenistic age and the contemporary medical man, who might often be a slave, contributed doubtless to the sterility of ancient Mediterranean medicine, including obstetrics and gynaecology. In the later Christian West there was not much incentive embryological study so long as the process of childbirth was left to the charms and incantations of barbarous midwives. But for a better insight into the economic position of embryologists in past ages nearly all the work requires to be done.

When working on his history of embryology, Joseph and Dorothy Needham got to know the distinguished historians, Charles and Dorothea Singer, as friends. They would spend much time, weekends and often whole weeks, with the Singers in London and at their manor-house home, Kilmarth, overlooking St Austell's Bay in Cornwall. Needham felt a filial relationship with Charles Singer, an unforgettable person, with whom he went swimming, collecting the delicious garfish, whose long spearlike snout made it unwanted, or attacking the overgrowth on the estate paths.

His great affection for Singer was reinforced by his experience when attending the Second International Congress of the History of Science and Technology, held in London from 29 June to 3 July 1931 in a lecture room of the Science Museum at South Kensington. In those days, the history of science was only of routine scholarly interest. As an academic study it was primarily an amiable diversion for retired scientists, or those about to retire. Only a handful throughout the world were engaged full-time in the subject. Thus, the International Congress was regarded as a routine gathering of historians and scientists, each pursuing his own field of interest with occasional attempts to correlate them.

However, the congress, at which Charles Singer presided, was transformed into 'a remarkable phenomenon' by the presence of a Soviet delegation, which arrived at the very last moment. It was an important group of eight men, certainly the most knowledgeable and responsible about the role of science in a Soviet socialist society. The leader was Nikolai Ivanovich Bukharin, a classical Marxist thinker and encyclopedist, the leading theoretician of the revolution after Lenin, co-leader with Stalin of the Party from 1925 to 1928, and head of the Communist International from 1926 to 1929. In Great Britain, he was known as an accepted decision-maker and spokesman for the scientific community as director of research under the Supreme Economic Council, and as a member of the Academy of Sciences, head of its Commission for the History of Knowledge. But he had been expelled from the Politburo in 1929, and his political fate as the symbol of the anti-Stalinist opposition was already decided: he was to be executed in 1938 following a show trial.

The other members of the Soviet delegation were the physicist A. F. Ioffe, the economist M. I. Rubinstein, the physiologist B. M. Zadovsky, the mathematician A. I. Kol'man,[4] the plant

4.  At the History of Science Congress in Warsaw in 1965, Needham met again with Kol'man, the only survivor of that 1931 delegation.

geneticist N. I. Vavilov, the electrical engineer V. F. Mitkevich, and the theoretical physicist B. M. Hessen. They had not been allocated any official time in the programme, but to meet their needs the congress was extended by a full half-day to hear their papers. That was not sufficient; they wanted to talk for hours, and Singer, who was chairman, used a large ship's bell to ring vigorously to shut them up after their twenty minutes each. However, to allow a full presentation, line by line, like fingers pointing a way in the darkness, translation of each paper into English was organized with back-breaking urgency, and within five days the necessary texts were available. These were assembled in the book *Science at the Crossroads*, and published, replete with grammatical and typographical errors, ten days later.

Although most of the participants at the congress rejected their Marxist approach, there were many who responded with great sympathy, particularly the group of young radicals present who included Bernal, Levy, Hogben and Needham himself. It was the paper by Hessen, 'The Social and Economic Roots of Newton's *Principia*', which startled them. Although Needham recalled it as 'perhaps the outstanding Russian contribution', it was not all that much of a surprise for the Needhams who had begun to read the Marxist classics in 1925. What was particularly interesting was Hessen's application of the ideas of historical materialism to the history of science.

Needham agreed with Bernal's statement on the two lines of argument discerned from the Soviet contributions.

The first demonstration was a historical analysis of actual discovery, particularly detailed in the case of Newton, showing the dependence of his thought firstly on the dominant technical problems of the day in navigation, ballistics and metallurgy, and secondly on the current political and religious controversy. Newton's work represents the scientific analogue of the Anglican compromise standing between the Aristotelianism of Rome and the rank materialism of Overton and his Levellers. Thus, even mathematics becomes in a sense permeated with political and economic influences.[5]

The Soviet papers did provide a seminal experience as they re-

5.   J. D. Bernal, *The Freedom of Necessity*, p. 337, London, Routledge & Kegan Paul, 1949.

vealed a road out from the sterile labyrinth that was then the history of science.[6]

As Cambridge historian, Robert Young, stated,

Reverting once again to Needham's Marxist thesis about men and their ideas being born of their time, it is important to add the corollary that changed men with changed consciousness are the product of changed times. This is as true of the present as it was of the nineteenth-century debate on man's place in nature. It was also true of Joseph Needham's development. In his essay on 'Metamorphoses of Scepticism' (1941), he reports that 'The process of socialization of my outlook, however, really began with the General Strike in 1926 [in which he was on the wrong side] and was completed by the rise to power of Hitlerite fascism in 1933'.

Between these two events there occurred the Second International Congress of the History of Science and Technology in London (1931), at which the Soviet delegation put forth the version of Marxist historiography of science which inspired the approaches of Needham and Bernal, and which influenced Crowther and others.

Needham later wrote in 'Limiting Factors in the History of Science':

In sum, we cannot dissociate scientific advances from the technical needs and processes of the time, and the economic structure in which all are embedded. . . . The history of science is not a mere succession of inexplicable geniuses, direct Promethean ambassadors to man from heaven. Whether a given fact would have got itself discovered by some other person than the historical discoverer had he not lived it is certainly profitless and probably meaningless to enquire. But scientific men, as Bukharin said, do not live in a vacuum; on the contrary, the directions of their interests are ever conditioned by the structure of the world they live in. Further historical research will enable us to do for the great embryologists what has been well done by Hessen for Isaac Newton.[7]

6.  As Needham stated in 1935 in his Carmalt Lecture at Yale University, published as, 'Limiting Factors in the History of Science, Observed in the History of Embryology', J. Needham, *Time, the Refreshing River*, pp. 144–5, London, Allen & Unwin, 1943.

7.  Robert Young, 'Man's Place in Nature, Changing Perspectives', in Mikuláš Teich and Robert Young (eds.), *Changing Perspectives in the History of Science*, Chap. XIX, London, Heinemann, 1973.

And in his 'New Foreword' (pp. viii–ix) in the 1971 reprint of *Science at the Cross-Roads*, Needham wrote:

Perhaps the outstanding Russian contribution was that of Boris Hessen, who made a long and classical statement of the Marxist historiography of science, taking as his subject of analysis Isaac Newton. . . . [It was] a veritable manifesto of the Marxist form of externalism in the history of science. . . . This essay, with all its unsophisticated bluntness, had a great influence during the subsequent forty years, an influence still perhaps not yet exhausted; hence its present reprinting is to be welcomed. . . . The trumpet-blast of Hessen may therefore still have great value in orienting the minds of younger scholars towards direction fruitful for historical analyses still to come, and may lead in the end to a deeper understanding of the mainsprings and hindrances of science in East and West, far more subtle and sophisticated than he himself could ever hope to be.

The Soviet paper reinforced Needham's realization that while the natural sciences were a wonderful activity of the human mind, their history and how they had developed through the ages was equally worthy of study. His profound conviction was that human life consisted of several irreducible forms or modes of experience. He made clear that it was meaningful to distinguish between the following forms – the metaphysical or philosophical, the scientific, the historical, the aesthetic and the religious, each being irreducible to any of the others, but all being capable of interpretation by each other, if in contradictory ways.

He recognized that none of these forms of experience could provide an access to absolute truth or was the only key that could unlock the secrets of the universe. He felt that there was a duty to experience them all in so far as one could, and that probably their contradictions could only be resolved in the act of living them all. Although he did not realize it at the time, he was expressing a true existentialist position. It explained the 'neo-mechanist' view which he advocated in the philosophy of biology: that is, that science had to be done in a certain kind of way, but it was not going to provide a key to absolute reality.

That was Needham's thinking in his undergraduate and research-student days, but as the years went by his need for a unitary view of the world became overwhelming. The first unitary factor he found was ethics. But its place in Collingwood's forms of experience always worried him. He could not allocate it to one or other of the five forms. The breakthrough occurred through politics, particularly Marxism. This was for Needham a remarkable revealing experience. He defined ethics as the rules whereby men may live together in society with the utmost harmony and the best opportunities for the development of their common good. And politics as the attempt to objectify the most advanced ethics in a structure of society, to enmesh the ideal ethical relations in the real world. Ethics and politics became the necessary cement to unify the divergent forms of human experience.

It revealed the mechanisms that had been at work throughout history and how Christian ethics could be incorporated in the society of the future. Later, but slowly, he came to feel that theology must adapt itself to what the sciences have to say about the nature of our galaxy and our universe, and that the sciences must not neglect the insights of the historian and the philosopher. He expressed these views in a remarkable series of essays and addresses on science, religion, philosophy and socialism, published in various books between 1925 and 1941. These include *The Sceptical Biologist*, a title derived from Robert Boyle; *The Great Amphibium*, whose Brownean title was taken from the incompatibility of the different forms of experience; *History Is On Our Side*, a contribution to political religion and scientific faith; and *Time, The Refreshing River*, whose title was a quotation from W. H. Auden, a poet he admired greatly.

These essays/reviews were not the product of a full-time professional writer, but rather the result of Needham's toils as by-products of evening reading or extraneous study while waiting for the completion of a distillation or incubation. They are to be regarded as a tribute to Needham's pertinacity.

Lunching with Needham in London during the early 1960s, he made me uncomfortably aware of how deeply he felt the rather schizophrenic-like behaviour imposed on him due to the inability to have funds made available for full-time history-of-science scholarship. As Robert Young points out, Needham continually broke new ground in the field of the history of science. Young wrote in his contribution:

My main purpose is to attempt to stimulate debate on requirements of a radical historiography of science in the current period – thirty-seven years after Needham laid out his position [an essay delivered at Yale in 1935] and forty-one years after the Soviet delegation to the Second International Congress of the History of Science and Technology came to London and dramatically introduced a version of Marxist historiography. . . . The attempt to work towards a radical historiography cannot at present be based upon a settled conception of what is meant by 'radical'. However, certain aspects of this conception can be stated. It is concerned with an approach to history which is critical and in the service of transcendence and liberation rather than mere reproduction, one which gives insight into possibilities for achieving a society which is not alienating and repressive.[8]

During the time Needham was a Bachelor of Arts and dined at the B.A.s' table at the college, the talk was much about introspective psychology. The impact of the works of Freud and Jung were making a serious impression on those concerned with complexes, neuroses, repressions, sublimations, the cathartic method and so on, allied with neuro-physiology and neuro-chemistry. Needham found these discussions of particular value to him in later years, when he himself was experiencing anxiety and obsessional neuroses that he said alarmed him considerably. When, in September 1992, I asked him to give me some instances of what had been alarming him, he replied that they had been of little significance, for he was able to explain them satisfactorily when they happened and so overcame them. He added, 'Such phenomena have always been frequent among great creative minds.'

That he was a creative person is without doubt. Like Francis Bacon in the seventeenth century, he 'could discuss Rainbows, Fiery Meteors, Fossils, Human Faculties, the Art of Metallurgy and the natures of Numbers'. Or, preferably, like Sir Thomas Browne, found nothing outside his subject. As Christopher Brooke, the historian of Gonville and Caius College, wrote, if they were 'to ask the question – with what names will the fellowship and the College of the late twentieth century be most clearly remembered in the history of learning and research, many of us, I think, would hazard the guess – as the College of Hawking and Needham'.[9]

Needham has never taken a narrow scientism view. In his

8.  Robert Young, 'The Historiography and Ideological Contexts of the Nineteenth-century Debate on *Man's Place in Nature*', in Christopher Brooke, *A History of Gonville and Caius College*, p. 343, Woodbridge (United Kingdom), Boydell Press, 1985.
9.  Ibid., p. 299.

early twenties, he recognized that 'the very rising tide of specialization has obscured the fact that there are not a few problems, especially in the fields of pure knowledge, which cannot be understood in terms of one subject'. There were many questions which urgently 'needed the synthesis of two or more illuminations'.

In the paper on 'Mechanistic Biology', which he contributed to *Science, Religion and Reality* (1925), a volume of invited essays he edited, he made it very clear that the scientific method was not competent to give formal descriptions of life, without involving, say, philosophy. His attitude was that there was no point in trying to hold 'a fundamentally antiquated faith. What we need is a faith in tune with all the most important things that we know about the solar system and about the universe that we live in.'

When, after the publication of Stephen Hawking's[10] highly successful *A Brief History of Time* in 1988, I asked Needham whether he had yet read it, he replied, 'Yes, and I've read it twice. It's an important work, especially in relationship to a theory unifying all of space and time, and the possibility of knowing the mind of God.' Needham had a high regard for his brilliant younger colleague. But he was devoutly religious, whereas Hawking was not. However, Needham was well aware of how world-views were shattered by scientific discovery, as those of the Middle Ages were by the Copernican revolution.

How he linked his religious belief to scientific development was made clear in an address at a celebration of Roman Catholic Mass in Caius College chapel on 13 March 1966. He was then Master of the College, and although the occasion had been popularized as the first occasion that a Catholic mass had been said in the chapel since the Reformation, he could not accept that. For him, the Anglican liturgy was as Catholic as that of the Roman Catholics. As he explained, together with millions of other Anglicans he was brought up to view the Church of England as both Catholic and Reformed; that meant it maintained the three ancient orders of clergy – the apostolic succession, the impor-

10.  Stephen Hawking, Lucasian Professor of Applied Mathematics, distinguished investigator of black holes in the universe, who is crippled by motor neurone disease.

tance of the sacraments and the necessity of interpreting the scriptures in accordance with the traditions of the Church. He was both a High Church intellectual and an intellectual who happened to be High Church.

When he first explained this to me I was somewhat astonished. He noticed this and pointed out that since the early 1970s he had been a 'roving' reader in Ely diocese, and, in that of Chelmsford, he was attached to the parish church of Our Lady, St John the Baptist and St Laurence at Thaxted.

As he told me on many occasions, for him the holy liturgy of the mass was the central action of Christian worship. It had resulted in essentially a communal institution with a communist moral, for all men and women were there as one, 'without any difference or inequality', sharing in the same loaf and the same cup. 'That this was a palpable manifestation of the creator of the galaxies in infinite empty space, the neutron stars, the red dwarfs and white dwarfs, the thousands, perhaps millions, of planets like our own supporting unimaginable humanities, is a thought that as we say "doesn't bear thinking about", yet this is the true measure of the majesty of this rite,' he said in his college chapel address. 'For me it typifies the trend of all evolution and all history.'

The trend was expressed in his conception of 'integrative levels'. This was basic to his thinking. He used to call that 'successions in time and envelopes in space'. If you took the envelopes first, it was obvious that different levels of organization occurred one within the other: ultimate particles, the proton and the electron build up atoms, and atoms build up molecules, and molecules build up large colloidal particles and cell-constituents and para-crystalline phases and organelles, and they in their turn are organized into living cells.

Above that level, the cells form tissues and organs, and the latter combine into the functioning living body, and human bodies (and those of some animals) form social communities. As the central nervous system becomes more complex so mental phenomena emerge until the incredibly elaborate psychological life of man is attained.

For Needham, the remarkable thing about our world was that the envelopes seem each to be analogous to past phases in the history of its development. There were inorganic molecules before there were living cells, and the origin of these depended upon the right environmental conditions to bring out the potentialities of the first protein molecules; and there were living cells before there were organs or tissues of metazoan higher ones, that is multicellular animals with bodies composed of differentiated tissues and a co-ordinating nervous system.

Primitive organisms existed before there were higher ones, and higher organisms before there were any social associations. Thus, the fundamental thread that seemed to run through the entire history of the world was a continuous rise in the level of organization. That was probably going on in other galaxies, other solar systems, where the conditions were just right, perhaps thousands or hundreds of thousands of times, for those things to emerge.

An important corollary was that any static or too conservative view of the present state of human institutions was quite impossible. Exemplified in such triumphs of living organization as the first evolution of the cell-membrane, the kidney-tubule, and notochord (a skeletal structure characteristic of all vertebrates and invertebrates), the flint knife, or the plough, the art of language and the skill of shipbuilding.

For Needham, human society would not always remain separated into national states with national sovereignties above the moral law and social classes with different privileges and manners. He had a conception of the co-operative commonwealth of all humankind which was dawning on the world. Our present condition of civilization was not the last masterpiece of universal organization, the highest form of order of which nature was capable. The transition from economic individualism to the common ownership of the world's productive resources by humankind would be a step similar in nature to the transition from lifeless proteins to the living cell, or from primitive savagery to the first community, so clear for him was the continuity between inorganic, biological and social order.

It followed that the future state of social justice was not to be seen as a fantastic Utopia, or as a desperate hope, but as a form of organization having the full force of evolution behind it. As exhibited in social affairs, the vast miseries caused by industrialization and modern warfare, the genocide perpetrated by the Nazis, inclined people to pessimistic conclusions, but we had to take a sufficient time-scale. Our civilization has existed for an exceedingly short space of time compared with the time taken in biological evolution. There were once in the world only autotrophic bacteria, that is, organisms that used inorganic material as the main source of carbon, but we can now see there will be some day be a co-operative commonwealth of humankind.

# 4

## Religion and socialism

*I believe in holiness because I experience it.*
*I don't view it as a personal presence,*
*but holiness is as vivid as sexual pleasure or hunger.*

Marge Piercy, *Body of Glass*, London, 1992

Needham never wavered in the conviction formed when he was a student that many of the greatest ills arose from a reliance on the natural sciences as the only legitimate means of understanding the universe. The person who had much influenced him in that view was Rudolf Otto, who maintained that religion was the irreducible 'sense of the holy'; it had to be considered as one of the characteristic forms of human experience, analogous to science, philosophy, history and the arts.

Another of his fundamental beliefs was that what was known of evolution – cosmical, inorganic, organic and sociological – had to be taken seriously by people of all religions. That led him to reflect on Henry Drummond's *Natural Law in the Spiritual World*, and to empathize with the work of the French Catholic and mystic, Teilhard de Chardin. A long time before he ever knew him personally or ever read anything he wrote, Needham had at the back of his mind some conception of what de Chardin came to call 'the cosmic Christ', that is an expression of the great human values that appear in all civilizations. He met him in Paris in 1947 and was impressed by his special personal quality, though he found his work at first difficult to understand. De Chardin, who had studied geology, had views on evolution which were regarded with great disfavour by the Jesuit authorities. Julian Huxley pointed out that the order in which de Chardin had taken his vows forbade the publication, and even the expression, of his views concerned with 'the reconciliation between scientific humanism and Catholic orthodoxy.'[1] He was 'banished' to

1. Julian Huxley, *Memories II*, p. 28, London, Allen & Unwin, 1973.

China, where he had time to express in writing his view that 'the only acceptable religion for man is one that will teach him first of all to recognize, love and passionately serve the universe of which he is the most important element'. He also wrote: 'The dream upon which human research obscurely feeds is fundamentally that of mastering . . . the ultimate energy of which all other energies are merely servants; and thus by grasping the very mainspring of evolution, seizing the tiller of the world.'[2] Only after his death, when his manuscripts were released for publication by his niece, did understanding come. Needham later became President of the Chardin Association of Great Britain.

When asked by left-wing bookseller, Dr Eva Skelley, to name a few books that had helped significantly to shape his thinking, for publication in the Golden Jubilee Catalogue of the well-known London radical booksellers, Collets,[3] Needham replied that he was faced with a personal historical problem. He had to begin with the statement that he was a 'lifelong socialist'. At the age of 17, he had a vivid recollection of shocking his very bourgeois father by greeting, with his great friend Frank Chambers, the Russian Revolution with open enthusiasm.

His hopes for a better postwar world appeared in a poem he wrote while still at Oundle.

> And when we proceed to the work of rebuilding
> The Civilization of Europe again
> We will cast off the trammels of ponderous ages
> And start right afresh letting no wrong remain.
> We will make a clean sweep of all errors and littleness
> Chase from our planet the spirit of Gain.
> We will make a new start – which shall not be in vain.

Of course, when asked in the 1980s for this statement for Collets, he could describe himself as a 'lifelong socialist', but his introduction to socialism began only in the early 1930s. As an undergraduate, the books he selected for the prizes he gained as

---

2. G. Magloire, 'Teilhard de Chardin tel que je l'ai connu', *Synthèse*, November 1957.

3. In late 1993, Collet's at 66 Charing Cross Road – known as 'the bomb shop' because of its alleged anarchist connections – became bankrupt and closed down. It left a tremendous gap in so far as the availability of left-wing literature was concerned.

a medical student were, much to the astonishment of his tutor, *Preces Privatae* by Lancelot Andrewes and *Serious Call to a Devout and Holy Life* by William Laws. He read, too, as much as he could of the books of two great Caius bishops, John Cosin and Jeremy Taylor, and the works of the Cambridge Neo-Platonists, especially Henry More.

His socialism was always distinctively Christian. Throughout his long life he continued to reject a distinction between his religious belief and his left-wing politics. Imbued with the universal message of Otto he was enabled to meet with complete sympathy Confucians, Taoists, Buddhists and Hindus during his years in China and other parts of Asia. He insisted that he could not have done that if he had been a purely materialist scientist of the old-fashioned, conventional type.

He was much influenced by the popular authors of his day, such social democrats as H. G. Wells and Bernard Shaw. Although he had expressed great sympathy for the Soviet Revolution, he had never read any of the Marxist classics, until after meeting in 1924, at the Marine Biological Station of Roscoff in Brittany, Louis Rapkine, a charismatic Lithuanian-Canadian biologist, just beginning his research career Needham recalled:

One must imagine the lovely summer weather, the boats rocking in the laboratory harbour, the Ile de Batz in the distance, and the scientific work going on inside the building. . . . That was the time when my wife and I first came to know Louis Rapkine, and to appreciate his extraordinary personality and character as well as his great scientific ability.[4]

He was the son of a Jewish cobbler who, to escape the pogroms, had settled in Canada. Rapkine had a distinct French accent which caused him to be mistaken for a native French-Canadian. He was French by adopted nationality, and did most of his scientific work in Paris. But the full range of his qualities came to be revealed following the Nazi occupation of France. He set himself – as science writer and administrator, J. G. Crowther, Director during the Second World War of the British Council Science Department, said – 'to save the flower of French science'. He succeeded, and in so doing for seven years gave up his own scientific research.

4.   Meeting in Memory of Louis Rapkine, Society for Visiting Scientists, London, 9 March 1949.

Rapkine's influence on the Needhams was largely ethical and political. For them he provided the first sustained encounter with someone who had experience of being poor, both as a youngster and as a student in Paris. His hero was Benedict Baruch Spinoza, the seventeenth-century Dutch-Jewish philosopher, who was expelled from the Jewish Community for his 'heretical views.'

Guided by him they began to read the great Marxist classics. They did their best with *Das Kapital*, but Needham's favourite was Engels' *Dialectics of Nature*, which he had read in German, in an edition published in Moscow in 1925. It did not appear in English until 1954. Their radical education was taking place against the background of the Science and Society Movement in England.

A key figure in this development in the 1920s and 1930s was J. D. Bernal. He was 'one of the few under-thirty-fives whose culture was general, was informed by the new and the developing, and was enriched by the old, which he could not evade because it was always audible to him. He was, as Hyman Levy once put it, "a sink of ubiquity".' For him, the scientist in capitalist countries was saddled with an ideology that separated 'science' from life in general.[5]

Needham was never a member of the Communist Party, but was much influenced by colleagues who were. During the 1926 General Strike he worked as a blackleg engine driver on the Great Eastern Railway. However, when the company tried to use 'the national emergency' to victimize the unionized railwaymen, he protested vigorously and led a walkout of the volunteers.

In 1931, the first Communist student group, a breakaway from the Cambridge Labour Club, appeared in Cambridge. It heralded a tremendous growth in radicalization: the Cambridge Socialist Club had 200 members in 1933 and almost 1,000 in 1938 in a university with less than 5,000 undergraduates. In the 1931 General Election the Needhams were active in the university and town branches of the Labour Party.

Bernal felt handicapped by the absence of an effective organization of scientists. Needham agreed, and together with a few like-minded radicals they began to be active in giving new meaning to the Association of Scientific Workers (AScW), a body

5.   Maurice Goldsmith, *Sage: A Life of J. D. Bernal*, London, Heinemann, 1980.

conceived in Cambridge and formed in 1918 as the National Union of Scientific Workers, but which had fallen into desuetude. It was not an organization of specialists based on one profession, and although not affiliated to the Trades Union Congress, it acted as a trade union ready to be of service to all scientists. For many years Needham was the AScW representative on the Cambridge Trades Council. His was a broad approach of social responsibility which radical scientists were then developing, especially in Cambridge, and this allowed him to improvise advice and proposals for activity.

He and Dorothy were involved actively with the Cambridge Scientists' Anti-War Group, whose special studies on air-raid shelters and protection against gas and chemical warfare were of importance as the Nazi threat became more apparent. In 1935, he was elected Cambridge Branch chairman of the Socialist League, a ginger group set up in 1932 by Sir Stafford Cripps[6] to stir up the Labour Party. How far Needham succeeded in developing an active branch is not clear. But in national terms the Socialist League itself did not succeed in making a determined impact on the Labour Party.

However, these links with similar thinking colleagues were of personal importance, for Needham had for many years been very isolated and treated 'as Ishmael', as he put it, in college society. When he became a Fellow, most of the older Fellows continued to cold-shoulder him. He was treated as 'an absolute outsider', as an individual who had no sympathy for their bourgeois conventions. Two exceptions were Stanley Cook, Regius Professor of Hebrew, and T. B. Wood, Professor of Agriculture.

During his Part II degree and early research years he was a member of the Cannula Club, which was made up of a very select group of advanced and research students. They included F. J. W. Roughton, who became professor of physical chemistry and a great investigator of the properties of haemoglobin; Malcolm Dixon, who was to write many books on enzyme action; and R. E. Tunnicliffe, who became a brilliant lecturer in biochemistry at Cambridge. The Club members met regularly for dinner, after which they would listen to a colleague reading a

6. Sir Stafford Cripps had a formidable reputation as a barrister and served in Labour governments as Chancellor of the Exchequer and Solicitor-General.

prepared paper, or talking on the research he was engaged in, or on what he was reading.

His membership of the Cannula Club was only one aspect of his innermost drive for a world-view, to which religion, science, philosophy and art contributed. Although in the Biochemical Lab he worked side by side with some of the outstanding intellects in Cambridge, he recalled he was the only one who 'had any use for liturgical religion, and I have been faithful to it all my life'.[7]

For Needham as a Christian, the nub was the doctrine of *Regnum Dei*. The Kingdom of God on Earth would assuredly come, but the time it would take would be the result of humankind's own actions. In practice, it would be socialism, which for him was inevitable, as well as good and right. Equally with his religious beliefs he has been faithful to left-wing socialism. But his form of socialism was essentially Christian, inspired by Conrad Noel and Jack Putterill at Thaxted. Therefore, he could never see eye to eye with the Communist Party. Karl Marx adopted what he regarded as Victorian rationalism, which became enshrined in Marxist tradition. Of course, he accepted that religion has often been enlisted on the side of the exploiters, but Jan Zizka's Hussites in the fifteenth century, or Oliver Cromwell's troopers in the seventeenth, or the peasant movements in Central and Latin America during this century, inspired by liberation theology, could not be seen as reactionary forces.

Needham first went to Thaxted to listen to Conrad Noel about 1927. He was still going there in 1993. His book, *Time, The Refreshing River*, was dedicated to Noel, 'Priest of Thaxted and Prophet of God's Kingdom on Earth'. Noel was an original, seeking to hold together the three themes of Catholicism, socialism and patriotism. He came from an aristocratic background. His full name was Conrad Le Despenser Roden Noel. His grandfather was an earl, and his father a minor courtier as Groom of the Privy Chamber.

Although Noel was a radical rebel, his name is associated with the parish of Thaxted in north Essex, where he was instituted in 1910, after accepting an offer from Frances Brook, Lady Warwick, to take over the largest of the churches under her patronage. During most of his years there, the patron of the living

7.  *China News*, Spring 1983, p. 248.

Joseph Needham, portrait by James Wood, *c.* 1963

Alicia Montgomery

Joseph Needham's father in Red Cross uniform

Joseph Needham as a baby with his mother

Joseph Needham, aged about 3

Dorothy Needham

Joseph and Dorothy Needham, outside the Biochemical Lab, Cambridge, *c.* 1928

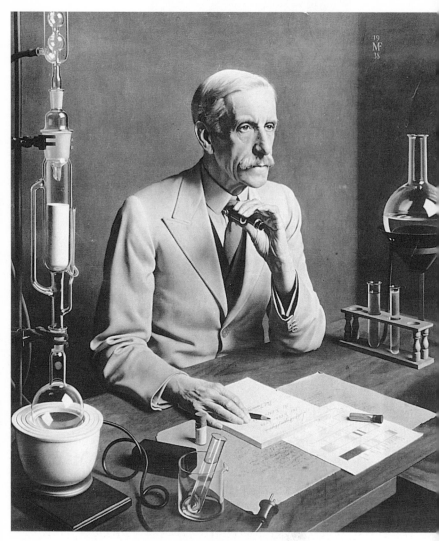

Sir Frederick Gowland Hopkins O.M., President of the Royal
Society, 1930–35, from the portrait by Meredith Frampton,
A.R.A. (Photo by courtesy of the Royal Society of London)

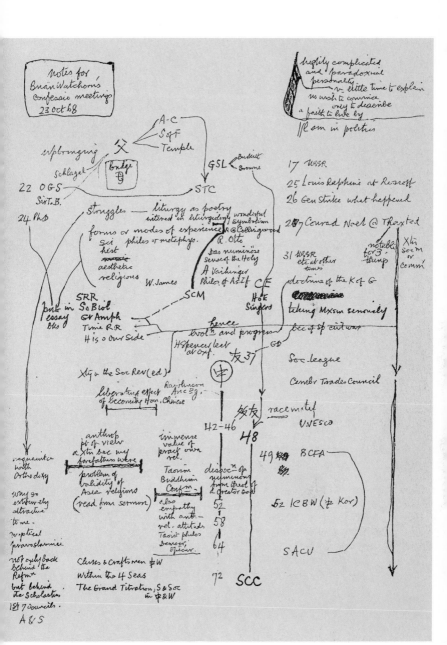

Notes for Confessio Meeting, 23 October 1968

Joseph Needham in Chinese dress, *c.* 1945

Lu Gwei-Djen and Joseph Needham, *c.* 1980. © *The Scotsman*

Joseph Needham, Dorothy Needham and Lu Gwei-Djen,
Hong Kong, *c.* 1980

Joseph Needham, mid-1980s

Joseph Needham in uniform

Joseph Needham and five Chinese colleagues in China, 1942–46

United Nations General Council, Mexico City

Joseph Needham, Caius, early 1950s

The picture from the Taoist Genii printed on
the cover of each volume of *Science and Civilisation
in China* is part of a painted temple scroll, recent
but traditional, given to Brian Harland in Szechuan
Province, 1946

'Needham: A Medal', *Cambridge Evening News*, 9 July 1994 (Photo: Dave Parfitt)

was the Countess. She was reputedly the most beautiful woman in the country, with a personal estate of 30,000 acres. She had been for many years a mistress of the Prince of Wales, later Edward VII. She became a socialist after a meeting with Robert Blatchford.[8]

It appears that following a luxurious Louis XVI ball she had organized, the *Clarion*, a weekly radical paper, featured a strong attack on the occasion. When Lady Warwick had this drawn to her attention, she hunted out the editor, Robert Blatchford, in his Fleet Street office. That meeting and subsequent reading converted her to socialism. She wrote to a friend, 'Socialism is the one religion that unites the human race all over the world, in the common cause of Humanity, and it is very, very wonderful, and it is growing as mushrooms grow, and nothing can stop it.'[9]

I went with Needham to Thaxted in July 1992 for mass and communion on the occasion of the fiftieth anniversary of Noel's death. He had a great influence on Needham, who accepted his view of the New Age to come on earth. It was to be a Kingdom of an International Commonwealth of all Peoples built on the Rock of Justice and Fraternity. In that kingdom, there would be no room for exploitation and inequality – no nation would domineer over others, none to be before or after, none greater or less than another.

At one time, Noel joined, as Lady Warwick had done, the Social Democratic Federation, whose leader was H. M. Hyndman. They proclaimed the existence of the class struggle rather than any brand of ethical socialism with its 'love thy neighbour' approach.[10] The federation was regarded as an exponent of militant socialism in the late 1880s when the phrase 'social democracy' was equated with Marxism and militancy.[11] Why, then did Noel join? Pepper said that it reflected 'his need for clarity and certainty at that time and his lack of sympathy for fudged and confused opinions. He was to say of himself later, "I

8.   A. J. Davies, *To Build A New Jerusalem*, p. 50, London, Michael Joseph, 1972.

9.   R. Groves, *Conrad Noel and the Thaxted Movement*, p. 24, London, Merlin Press, 1967. (Cited by Leonard Pepper, *Conrad Noel*, London, Church Literature Association, 1983.)

10.  Davies, op. cit., p. 47.

11.  Ibid., p. 46.

was never an undogmatic undenominationist in religion or politics." He had passed beyond the stage of adopting socialism as an inspiring humanitarian creed.'[12] Although he was involved with the SDF for a short period only, his Christian socialism throughout his life continued to have an unmistakable Marxist approach to political and economic problems.

Surprisingly, Noel was not opposed to the First World War: he thought it right to oppose German militarism. He had a collection of flags of the Allies placed in the church to which in 1916, after the Irish uprising, he added the tricolour of Sinn Fein, and in 1917, after the Russian Revolution, the red flag of the Communist International inscribed with the words, 'He hath made of one blood all nations.' When the war ended in 1918, he continued to display those two flags plus the flag of St George. For him the Union Jack was not 'the old flag of this country. It is the modern flag of brute force domination.' This led to 'the Battle of the Flags'. It was during the miners' strike of 1921 that someone removed the red flag. Noel replaced it. Again it was pulled down by Cambridge students and young army officers. Noel put up another higher in the church and had it guarded night and day. Some local inhabitants sought but failed legally to have the flags removed. The Church Council backed their rector. However, in 1922 a consistory court ruled that the flags had to be removed. From the pulpit Noel declared, 'The flag has been removed but the preaching will go on.'

Noel and Jack Putterill, his successor as vicar, continued to express a militant Christian socialism, which much influenced Needham. He was much affected, also, by their introduction to him of an understanding of the relationship between folk dancing and religion. Folk dancing was brought to Thaxted by Noel's wife Miriam in 1910. Noel involved Mary Neale and her Esperance Club, 'which brought Thaxted into direct relationship with a traditional rural culture which had survived industrialization, and was finding an unexpected resurgence of life'.[13]

There were some outstanding local dancers, but activities stopped with the First World War. The revival came after the

---

12. Leonard Pepper, *Conrad Noel*, p. 9, London, The Church Literature Association, 1983 (Oxford Prophets, 16) .
13. Groves, op. cit., p. 70.

war along lines advocated by Cecil Sharp, founder of the English Folk Dance and Song Society and a socialist. The morris dance is the oldest unchanged dance in England. Its origin is Moorish, hence 'morris'. It is said to have been introduced by John of Gaunt on his return from Spain.

All through the 1920s and 1930s, Needham was a very keen morris dancer. There was a clerihew that went:

> Doctor Joseph Needham,
> Dances with philosophical freedom.
> You must mind your toes if
> You chance to dance with Joseph

He was attracted particularly because it is associated with May Day, the day of people's liberation, and because it was in many regards a late survival of pre-Christian festivals. It was for him the authentic expression of the voice of the oppressed. Linked as it was with seasonal festivals it represented 'a means of escape from the state of subjection in which the peasant lived'.[14] The bourgeois puritan and businessman hated the morris and the maypole, because they 'wanted to make the world safe for the profit of godly industry'. The morris dance linked up with those wide, democratic, socialist and populist elements which determined his lifelong political outlook. 'In the socialist state of the future, as in the Soviet Union,' he wrote, 'the traditional ritual dances of the people will be treasured indeed. They are the pure creations of the working class and they will unite by a remarkable continuity the developed communism of the future with the antique primitive communism of the past.'[15]

Needham was familiar with the traditional dances of England – the morris of the south-west, and the long-sword and short-sword (rapper) dances of the north-east. He belonged to the Cambridge Morris Men, who combined to dance with the traditional dancers of Chipping Camden in Oxfordshire and the lead miners of Winster in Derbyshire. A fellow dancer was C. H. Waddington, his close collaborator in experimental morphology and biochemistry. With him and Arthur Peck, a Greek scholar at Christ's College, and with others, they set up the

14. Joseph Needham, *Time, The Refreshing River*, p. 130, London, Allen & Unwin, 1943.
15. Ibid., p. 130.

that he began to learn their language, difficult for Europeans inasmuch as its written form is ideographic and not alphabetic. They helped him to acquire an understanding of the rudiments and he began to practise it in a series of letters to them.

Of course, eventually, he mastered the language, though, as he later 'boasted', he did not know a single Chinese character before he was 37 years of age. Suddenly, he found a Chinese dictionary most exciting, and he spent a great deal of his spare evening time constructing his own dictionary. But the person who gave him much immediate help was the Czech scholar, Gustave Haloun, Professor of Chinese at Cambridge, who understood that ordinary linguistic teaching was out of the question for a busy, almost-40-year-old scientist, passionately determined to learn the language. He invited Needham to spend a few hours each week going over his translation of the difficult philosophical and economic fourth/fifth century text, *Kuan Tzu*, which he was preparing for publication. It was a delightful, exciting introduction to ancient Chinese philosophy and the language used.

Needham learned his Chinese as a labour of love, which stimulated him to invent his own ways of mastering the language. In the various notebooks he used, he always set apart a page for each consonantly termination, such as *-ien* or *-iang*, with four columns, each for a tone in the national pronunciation, and then on the left a succession of consonantly initials, such as *ch-, ch'-, f-, j-,* and so on. Thus he produced a series of matrix tables like graphs, with the meanings of the monosyllable inserted in columns. This was helpful in memorizing words.

He also made up a dictionary based on a principle he invented. This approach divided all characters into four main classes, with a small addendum for 'miscellaneous'. The classes were strokes that went straight down, veered off to the right, swerved to the left, and finally enclosures (such as *k'ou* and *hui*.) All the classical 214 radicals of the lexicographers could then be subsumed, with all their derivatives, under one or other of the four major divisions.

He devoted a third notebook to entries concerning 'configurations', which in the West is called 'grammar', but which some have supposed Chinese does not have. Needham pored over word order, classifiers, 'empty' particles, numerals, conjunctions and punctuation. He became quite fluent and enjoyed reading a page of Chinese: 'It was like going for a swim on a very hot day, because it got me completely out of the alphabet.'[1]

But a question to which he could find no convincing answer was posed by the fact that the more he got to know the three Chinese, the more he was puzzled by why modern science had arisen in Europe alone. His fundamental answer to that question is provided in over a half-century of devoted effort, which would not have been initiated or made possible without his special relationship with Lu Gwei-Djen. She was the dominant influence that caused him to change from being a regular research scientist – a biochemist, specializing in chemistry and embryology – to become a historian of science in China. Her father, Lu Mao-Thing (Ding), was a distinguished Nanking pharmacist, expert in both traditional Chinese and modern *materia medica*. He had ensured that she was brought up both to understand modern science and to appreciate the achievements of the ancient Chinese practitioners and artisans.

Before she left Cambridge in 1939 to do research in the United States – first, at the University of California, Berkeley, which she had to leave because of an allergy picked up from the flowering acacia tree, then at Birmingham City Hospital, Alabama, and the College of Physicians and Surgeons, Columbia University, New York – she worked out with Needham the subjects for a book to seek to provide an answer to the great puzzle worrying him. Neither could foresee that in time they would give birth to a work so creative and original, *Science and Civilisation in China (SCC)*, that the body of historiography would be changed.

It was China itself that reinforced Lu Gwei-Djen's influence, for early in 1942 the British Government invited Needham to go to China, then under invasion and partial occupation by Japan, as a representative of the Royal Society to extend Anglo-Chinese relations in the cultural-scientific field. 'I was to do everything in my power to renew and extend the cultural bonds between the British and Chinese peoples.' It was then that Needham's 'second half-life' became a reality. Amusingly, he

1. For a detailed statement on Needham's approach see, in particular, Chapter 2: 'Conventions and Abbreviations, Romanization of Chinese Characters, Note on the Chinese Language', *Science and Civilisation in China*, Vol. 1, pp. 18–41.

comments on his recommendation by the Royal Society: 'I was probably the only Fellow who could speak Chinese and with my reputation I was regarded as dispensable.' At the same time E. R. Dodds, Oxford Greek philosopher, went out to represent the British Academy. They made up a mission dreamed up by Sir George Sansom, of the Diplomatic Service and well-known expert on Japan.

Needham did not go directly to 'Free China'. First, he made official and unofficial contacts in the United Kingdom, the United States and India with organizations concerned with science. He had already begun to formulate what were to be his tasks, and the visits he was making were to let the organizations know what he was proposing to do and to ascertain how they could help him. In London, for instance, he went to the Chinese, American and Soviet Embassies, and to the Chief Scientific Adviser to the Admiralty, the Scientific Advisory Committees of the Cabinet and of the Ministry of Production, the Scientific Advisers and Intelligence at the War Office. Communication would be of importance, and he had talks with the Far East sections of the BBC and the Ministry of Information.

The thoroughness with which he did this in all three countries was typical of his organizational ability. He believed that the apparently banal functions of a 'post-office' serving science and learning would be most helpful. He remembered the 'post-office' run by Oldenburg, the first Foreign Secretary of the Royal Society, who, with his wide European contacts, had received letters from the microscopist Leeuwenhoek describing for the first time the 'new world' beyond the limits of ordinary vision. Needham was to describe for the first time an 'old world' of science and engineering and medicine, and to eradicate the traditional limits of social vision.

When, in February 1943, he arrived by plane at Kunming in Yunnan, he surprised himself and the British Consul-General, A. G. Ogden, in being able to talk with Chinese people fairly freely. The flight from Calcutta was tiring, but he was too excited to sleep. He tried to compose a poem, but he felt too drained of oxygen. Instead, he had, as always, a volume of Lucretius with him, and he tried to improve on Leonard's English translation. At the airport he was met by the Vice-Consul David Hough, the King's Messenger Pratt, and Colonel Yen, dressed in black gown, representing the Chinese authorities.

His first impressions of China are vividly described in a letter to Margaret Mead, the American social anthropologist, who had made him promise to write of his first thirty-six hours in the country. He was struck by

the prevailing colour blue of sky and blue of civilians' gowns and peasants' coats and trousers, and yellow of earth. A very nice yellow, warm like Cotswold stone. . . . Entered the town Kunming through a gate, general impression not unclean, streets paved but sidewalks just earth mostly. Houses two-storey. . . . All faintly reminiscent of an English village street. . . . Weather almost exactly like Cambridge in spring or autumn, and with innumerable rooks cawing, one might think oneself at Duxford vicarage if one closed one's eyes. . . . As I write there are many patches of blue sky. Everything seems so strangely familiar (having thought so much about China for so long) yet like a dream. . . . The Consulate and most of the houses I've seen are all slightly hick, reminding me of Kingstead Mill, as if one should have to live permanently in the country. . . . The Suifi Lake close by is quite reminiscent of Coe Fen at Cambridge. . . . Another resemblance to Cambridge is the continual noise of planes.[2]

That letter with its nostalgic references to Cambridge was written on 26 February 1943. On 1 March, he wrote again from the buildings of the Chinghua University Research Institute, near the village of Tapuchi, nestling in the mountains about as far from Kunming as Barton is from Cambridge. He found 'the labs, all mixed up with the living rooms, built of mudbrick, plastered, and in the form of quadrangles. I have a bedroom and a sitting-room, but hardly anything in them save bed, table and chairs. People like being here because it's so safe from air raids, though there haven't been any here for over a year, Kunming being now well defended by the US Air Force.'

He was delighted to find that 'when introduced to a new group of Chinese academic people one can absolutely *depend* on their being charming people. . . . This could be a place where I might settle down, if I had a jeep, and could get some rugs or matting . . . but it would be very far from the centre of Government. Until I've spoken with the Chinese Government and the Ambassador, I can't have much idea of where I must fix my abode.'

2.   J. Needham and D. M. Needham, *Science Outpost*, p. 27, London, Pilot Press, 1948.

On 21 March 1943, he was en route by plane to Chungking, where as Director of the British Scientific Mission in China, he was welcomed by the Ambassador, Sir Horace Seymour, at whose villa he met Sir Eric Teichman, an Old Caian. A month later, despite communication difficulties – phone, car and language – he was saying farewell to Dr E. R. Dodds, who had arrived somewhat before him, only to find that the Humanities were not regarded as vital to the war effort. Needham was discovering that on the scientific and technical side there was everything to be done. Scientists and engineers required much practical help to carry out their work under unbelievably difficult conditions.

Although he found everyone highly co-operative, he perceived there were submerged rocks and snags. He recognized that 'public work is like entering a power-house switchboard hall. There are a thousand switches. Some of them, you know will do what you want to get done, but there are no labels. You proceed to push one button after another. Usually, nothing happens, but every twentieth one sets something in motion according to your desired line. To avoid waste of time, you have to develop some sixth sense to know which will be the most effective buttons to push.' Needham was having to learn fast. He wanted to get the blessing of the Chinese Government to construct a Sino-Western Science Co-operation Office, which would outlast the war and prove to be a basic piece of international machinery. That was paramount, but at the same time there were a hundred other jobs, such as organization of an urgently necessary supply service of essential chemicals and research apparatus from India, and lecturing to students and scientists.

The Sino-British Science Co-operation Office (given that name in deference to the host country) that Needham set up was financed by the British Council, and supported vigorously by J. G. Crowther, director of the Council's Science Department, and under the aegis of the British Ministry of Production. It was part of the Allied attempt to break the Japanese intellectual and technical blockade. Those in the S-BSCO regarded themselves not as an outpost of science because they were in China, but as an outpost in western China. With the Chinese army they had their backs to the wall: behind them was the Tibet massif and the Gobi Desert. Around them, for Chinese scientists from the great eastern cities, were the most primitive conditions. For example, at Chiating nuclear physics was discussed in a family temple; in the caves of Kuangsi there were large power stations and engineers dying to talk to a technologist from the outside world; and among the aboriginal tribesfolk of Tali a planktonologist was helped to launch his boat. Needham saw them all as 'manning a series of outposts' in which there was an extraordinary mingling of the old and the new.

The Science Cooperation Bureau was the clearing-house for scientific interchange. The work was done under the following headings:

1. Supply of already existing information on problems arising in China: (a) pure science; (b) applied science.
2. Supply of ideas from the West to meet specific problems arising in China: (a) pure science; (b) applied science.
3. Supply of essential research chemicals and apparatus: (a) pure science; (b) applied science, fine scientific production; (c) practical teaching.
4. Supply of scientific literature: (a) microfilms; (b) reading machines; (c) journals; (d) diminutives; (e) offprints; (f) typescript copies; (g) books.
5. Supply of manuscript scientific papers, articles and reviews from the West for publication in Chinese scientific journals.
6. Special services, such as an offer to cut thin rock sections by the Indian Geological Survey; printing of maps for the Chinese Geological Survey by the Indian Cartographic Office; and provision of list of edible and poisonous plants, in North Burma and the Shan States, for the Surgeon-General's Department.
7. Contact with the West: (a) transmission of important scientific correspondence between Chinese and Western scientists; (b) contact with Chinese scientific mission in London and Washington (if appointed); (c) contact with government departments concerned with science in London and Washington.
8. Output of scientific material from China: (a) manuscript papers by Chinese scientists for publication in Western scientific journals; (b) current Chinese printed, duplicated or mimeographed publications for Western scientific libraries; (c) translation from Chinese of good scientific papers for publication in the West; (d) abstracting service for Western abstract journals; (e) *Acta Brevia Sinensia*, science newsletter prepared by the Natural Science Society of China.

9. Output of technical memoranda on matters arising in China for the help of the other United Nations.
10. Output of information about Chinese science for publication in the West: (a) articles by Chinese scientists and this office; (b) photographs of the scientific side of the Chinese war effort.
11. Output of fine actual scientific products from China.
12. Advice to the Chinese Government, if desired, at any time, on possible new industries, etc.
13. Assistance with problems of exchange of personnel: (a) mature students, (i) industry, (ii) university; (b) technical experts; (c) research and graduate students.

The above programme was outlined by Needham before his arrival in China, and to ensure the utmost efficiency in scientific exchanges through the widest possible range of contacts on both sides he had made his personal visits in the United Kingdom, the United States and India.

Finally, Needham was able to secure agreement that his Science Co-operation Bureau would be associated with the Council for the Promotion of Science in the National Defence – *Kuofang Kohsueh Chishu Tsouchinhui (Guofang Gexue Jishu Zoujinhui)* – with its five divisions concerned with manufactures and technology, communications and medicine, scientific personnel, cultural contacts and publicity, miscellaneous affairs. The bureau was also able to make arrangements to have the advantage of named liaison officers within the principal Chinese ministries – War, Economics, Education, Health, Agriculture, and Academia Sinica.

Needham's workload was enormous. In his report in February 1944 on the first year's working of the bureau, he gave details of his visits to universities, industrial plants, research laboratories and arsenals, each visit including a large amount of lecturing. He went everywhere where anything of importance was going on in science and technology, wherever the Japanese had not penetrated. He made journeys within China of about 8,000 km, mostly by motor truck, but also by plane, river steamer, junk, sampan, skin-raft, horseback, jeep bus and bicycle. He was accompanied to begin with by Huang Hsing-Tsung (Huang Xing Zong), a young organic chemist from Hong Kong University who, if not for his wartime work with the bureau, would have gone to Oxford as a Rhodes Scholar. His association with Needham was for just a little more than a year during 1943/44.

Huang Hsing-Tsung states that his association was

intensely eventful and absolutely unforgettable. I had the unique privilege of observing Needham in action at close range, as secretary, aide-de-camp, interpreter, time keeper, travelling companion and collaborator. We shared the rigours of travel in wartime China, all the way to the Northwest along the Old Silk Road to Tunhuang [Dunhuang] beyond the Great Wall, and through the Southeast over lush hills and vales to Foochow [Fuchou] by the East China sea. We shared moments of exhilaration, frustration, danger, relief, joy and despair.[3]

He was able to witness how Needham reacted to the reality of a China with its back to the wall, exhausted after years of external aggression and internal strife, a far cry from the romantic attachment of Cambridge in the 1930s, and how China in turn responded to him.

Huang Hsing-Tsung first met Needham on 1 May 1943. He had come in response to a letter from Needham made impressive by the prominent seals and stamps on the envelope and the letter itself. Needham wrote that he needed a secretary. Huang had been recommended highly by Professor Gordon King of Shanghai Medical College at Koloshan (Geloshan) in Chungking (Chongqing).

It was early on that cool, misty morning when he rang the bell at the home of Professor Ho Wen-Chun, near the campus of West China Union University, outside the city of Chengtu, the capital of Szechuan province. Mrs Ho showed him into a study and about ten minutes later 'Needham appeared. Wearing a loose blue Chinese gown, with his hair slightly dishevelled, he loomed large and foreboding, but his manner softened as he started to speak. I introduced myself. He took a blank card from the desk and proceeded to write my name in Chinese. After a couple of starts he wrote down all three characters correctly.' Needham looked in his diary, and they arranged to meet the following morning.

When they met again, Needham looked less formidable wearing a well-fitting army khaki shirt and shorts. He had two

3.   Huang Hsing-Tsung, *Peregrinations with Joseph Needham in China, 1943–44* (unpublished manuscript).

visitors, and Huang watched him complete 'the ubiquitous cards' with the full name, any courtesy or pen names, professional affiliation, scientific discipline and other interesting bits of information. He wrote the latter down in Chinese or English, or Latin, Greek, German or French. The cards grew into an extensive registry of scientists in China.

Huang was offered the job, which he began on 17 May. He was impressed by the fact that Needham had chosen to stay with a Chinese family, rather than with a Western family on the campus in a spacious house with modern comforts. It indicated to him that Needham was mentally attuned and physically ready to endure the discomforts of China, to mingle with the people in their own modest surroundings.

When, on 26 May, Huang arrived at 7 a.m. to leave with Needham on a long trip, he found Needham had a swollen arm arising from a tetanus injection he had had a couple of days before. But the doctor pronounced him fit to travel. They left in the Wuhan University car at about 9.30 a.m. and arrived at Loshan at 5.30 p.m. They made many visits during the next five days. Needham liked the symbolism of the library and the Colleges of Art and Law at Wuhan University being housed in a Confucian temple. Needham gave several lectures. At the end of one at the Polytechnic, without warning he delivered a short speech in Chinese. Huang was 'pleasantly surprised since I was able to understand him quite well. Up to that point, I had only heard him speak a few words of conversational Chinese.'

Within a few days, Huang was to hear Needham lecturing in German. That was at Lichuang (Lizhaung), which could be reached only by boat. With them was Professor Shih Seng-Han (Shi Shenghan) of Wuhan, 'a very Cambridgish plant physiologist', who had arranged for them to travel on a salt transport boat down the Min river to Ipin, where they could catch a passenger steamer to Lichuang. As Needham noted on 3 June, 'You'd never believe it, but I am on a "junk" travelling down river!' The journey had nightmarish moments. They had to transfer to a smaller boat, in which they rocked all day, so that by the time they got to Lichuang, Needham was swaying to and fro. They had spent a miserable night in a central cabin so small, with a roof so low, that they could scarcely lie down or sit up with any comfort. They sat in a crouched position most of the day as it was raining hard.

When, at about 5 p.m., they arrived, they were greeted warmly by a group of professors from the German-Chinese University of Tongchi. They were put up in a new, simple guest house with a caretaker and a cook to look after them. Bread was available and Needham enjoyed his breakfast toast. When, two days later, Professor Shih (Shi) left them, he presented Needham with a scroll on which was inscribed, '*Rjêng tao ming cheng, Ti tao ming shu (Jeng dao ming zheng, Di dao ming shu)*'.[4] Needham turned this into a Popian couplet: 'Nature from growing trees we best discern/And man's estate from social order learn.' That is, that human society is as natural as any of the lower-level biological phenomena.

Needham was in good form. One morning, officially Engineers' Day, he conversed and lectured in German at one of the most inspiring meetings he said he'd ever been at. The university administrative centre was in an ancient temple dedicated to Ta (Da) Yu, the legendary, first waterworks engineer. While listening to the speaker who preceded him, Needham studied the magnificent gilded carvings on the roof overhead, for him reminiscent of Thaxted. He was well aware of the importance of the waterworks for all Chinese history. He spoke 'first of true German culture, as opposed to Nazism, and then about universities and students in England and America, and finally about why science didn't develop in China, bringing in waterworks, of course. He spoke for an hour and you could have heard a pin drop.'

Huang relates that Needham spoke so much German that he would sometimes speak to him in German when they were alone in the guest house.

One evening, he was in a rather playful mood. He asked me a question and I responded in German. He stood up, stretched out his right arm in a 'grand gesture' and made a statement in German. He then abruptly sat down, allowing the full weight of his body to drop all at once on the flimsy rattan chair. One of its legs gave way with a loud crack. With great agility he leaned forward and averted a disaster. I said that this was a warning to him not to throw his weight around. He laughed, I laughed, and we had a good laugh together.

4.  The quotation was taken directly as it appeared in the book, *Science Outpost*. But Needham agreed that the word '*Rjêng*' should be replaced by '*Rên*'.

It was at that time that Needham wrote that Huang 'is a charming companion as well as most efficient. He should learn a lot of general science in this job.' He was to end up in charge of the Biological Programme of the United States National Science Foundation.

About four miles away was the Academia Sinica Historical Institute, with a staff of sixty, the biggest of the academy's institutes. A group of manor houses contained departments of history, archaeology, linguistics and physical anthropology. Needham was excited by his contact with probably the greatest concentration of intellectual power devoted to the study of Chinese civilization. He wrote:

My numerous inquiries about History of Science problems caused a general stir and various members of the Institute were running about digging out interesting stuff they'd come across, e.g. passages about firecrackers in 2nd Century AD: accounts of great explosions and decrees forbidding the sale of gunpowder to the Tartars in 1076 AD, i.e. 2 centuries before Berthold Schwartz's alleged original discovery in the West.

He was also highly impressed by the archaeological treasures kept in humble buildings. Things he had read about and now saw for himself: Chou and Han bronzes, Shang oracle bones, and bamboo tablets on which the ancient classics were written.

He and Huang stayed one night with the great scholar, Fu Ssu-Nien (Sinien), Director of the Historical Institute. He was somewhat Westernized and a fascinating talker. It was at that institute that Needham first met Wang Ling, an assistant research fellow, who, a few years later, was to become his first collaborator at Cambridge in the *SCC* project. His discussions with local scholars reinforced his suspicion that there existed countless material on the history of science and technology which had to be found and made available to scholars in other lands.

For Needham, China was 'an extraordinary place'. The colour, movement and variety were so striking. China was not, as most Westerners believed, purely agricultural and artistic. The Chinese were the first discoverers of the magnetic compass, gunpowder, paper and printing, and the earliest technique of vaccination. In recent years, as he was witness to, modern science was growing there very strongly.

For instance, at Luhsien (Luxlon), about 120 miles down

river from Lichuang, where he visited the Twenty-third Arsenal, he left a very dirty boat, full of country folk spitting in all directions with high velocity, in cold and rainy weather, to re-embark six-and-a-half hours later on an elegant Arsenal boat. The guest-house had running hot and cold water, a functional WC, comfortable beds and a large sitting room. He met nothing like it anywhere else in China. And that was topped by 'a 100% European dinner including butter and peach tart – most extraordinary' at the Director's house.

For two 'amazing' days, they visited the chemical defence Arsenal, which was producing large quantities of bleaching powder. Some 17 miles away was the Research Institute. 'There is *no road*, one walks or is carried in sedan chair over coiling paved paths through the rice fields or up and down the hillsides.' Arrival was at an enormous place built between a cliff and a river, utilizing a natural cavern containing a three-storey laboratory and library. There were many other labs and pilot plants around. Needham regarded it as a miracle that they had moved everything there and built it all in one year.

Lunch included ice-cream, unique in the middle of Szechuan. Needham gave a lecture to the staff, and made preparations to leave before dawn, at 3.30, for Chungking on the Yangtze river steamer. Huang noted that because Needham was in a khaki uniform, the sentries they passed would salute him, and he would salute them back smartly and with great dignity.

Settling in in Chungking was not proving easy. The Ambassador, Sir Horace Seymour, and colleagues were sympathetic and helpful, but Needham found that his fundamental trouble was that 'I am my own little unit'. He had no resources at his command: for instance, he was always begging other people for card folders, paper clips, etc., unobtainable in the city. And what was much worse was that he had nowhere to live. 'Of course, I can't make a real fuss, because of these numerous tours on which I have to go, and one can put up with even Victory House (a hostel for visiting foreign dignitaries) for a week or two.'

On 7 August, he began his north-western trip in a couple of two-ton Chevrolet trucks on each of which Needham had the inscription 'Sino-British Science Co-operation Office' painted in bold Chinese characters on both sides. In Truck 1, there was Sir Eric Teichman (Tai as he was called), on his way to the Consulate at Tihua (Dihua), Sinkiang (Sinjiang), plus two mechanic-drivers and one servant. In Truck 2, with Needham, there were Huang Hsing-Tsung (Xing Zong); an American oil geologist, Dr Edward Beltz; Robert Payne, poet and writer of Futan University; and mechanic-driver, Kuang Wei (Guang Woi).

Needham wrote later, 'I had simply no idea of the sort of thing it [the trip] would be. Great mountain passes, overwhelming scenery, unpredictable roads, bridges broken down, roads washed away, truck traffic carrying on in spite of everything, and strange places to sleep in night after night.'

Needham was to be away for over four months, and Huang even longer – five months, two weeks. They travelled in the provinces of Shensi and Kansu (Gansu), whose southern part is loess country in which every geological feature has a layer, often hundreds of feet thick, of sand and dust particles from the Gobi Desert, deposited by the wind over centuries. The soil erodes easily to provide extraordinary canyons with vertical walls.

Needham's truck kept breaking down with gasket trouble. They had no spares. Teichman had taken them with him and gone ahead. Typical of the difficult situations they found themselves in was at Wukuanho (Wuganho), where the road had been washed away and there was a disorganized line of trucks and mule carts blocking a single-lane, temporary road made of boulders and pebbles. They had to pass more than twenty-four hours there, and when they left they had to carry on with a broken left rear spring, damaged by the boulders they had to go over. Arrangements were made for repairs to be done at Shuangshipu (Shuangshibu), in a technical school maintained by the Chinese Industrial Cooperatives (CIC). There were a number of such technical education establishments, known as Baillie schools after a pioneer ex-missionary.

On 18 August 1943, Needham drove his converted ambulance, his flagship, into Shuangshipu, a kind of junction or three-way village on the road north from Chengtu, in Szechuan (Sichuan), to Lanchow (Lanzhev), in Kansu. Because of its key position for communications, it had become a little town with various depots and small factories. There he met the fabulous New Zealander, Rewi Alley, of the CIC, of whom Needham said he never met a better friend and a more reliable colleague. In his house, partly a cave in the loess cliff, Alley gave them 'a wonderful meal'. They stayed in the CIC hostel for two nights, on one of which some Baillie boys sang folk and guerilla songs.

Needham joined in with some English folk songs and a demonstration of a morris dance.

The CIC was an organization set up as the Chinese withdrew from the eastern provinces and the great cities of the eastern seaboard overrun by the Japanese, and settled into the less-developed western provinces. They were 'trading space for time'. However, although the disorganization was great, the CIC made it possible for craftsmen, in particular, to group and organize themselves in producers' co-operatives, able to earn a livelihood and to contribute to the war effort. Through Baillie schools, hundreds of boys and young men, and some girls, were given much needed technical training.

Alley was looking for a place to set up a new Baillie school, so Needham was delighted that he agreed to join forces and travel together along the Old Silk Road towards the borders of Singkiang (Singjiang). With him went two boys from a Baillie school, who spoke the Kansu dialect. Although they were to have a most hectic time, Needham had what he termed 'the inestimable privilege of living at close quarters with Alley for several months. My education in Chinese ways and recent Chinese history was thus placed on a singularly firm foundation.' He described Alley as

a tough New Zealander, fair haired and ruddy, with good physique damaged somewhat by malaria caught in the south, he combines the ideas of Conrad Noel and J. B. S. Haldane in a personality not unlike that of Stanley Wilson. With a vast knowledge of many dialects of Chinese, customs, folklore, etc., he has an extraordinary capacity for human contacts with the most rustic villagers.

Though inured to much travel under the most uncomfortable circumstances, he is a scholar (in Shanghai he had a fine library). He went to Shanghai via Australia, he became chief factory inspector there. . . . Eventually became the founder, spark-plug and mainstay of the CIC.

Their stay together on the journey was to be for several months. That had not been intended. What happened was that after visiting the oilfield at Lao-chun-miao, and after Alley had decided to site the new Baillie school at Shantan (Shandan), they visited the famous cave-temples of Tunhuang (Dunhuang) at the oasis of Chienfotung, some 25 miles across the desert to the south of the city. There the truck's big end snapped. The engine was sent to the oilfield some 200 miles east on a bullock-cart to be repaired. Needham, Alley and the boys camped out most uncomfortably by the cave-temples.

For Needham that was an unparalleled opportunity not only to learn more from Alley and the boys, but also to study those thousand years of Buddhist art embodied in the acres of frescoes, painted walls and ceilings of the temples. When the engine came back and they were entering dilapidated Tunhuang (Dunghuang), Alley, who was riding on the running-board, shouted, 'Now we're going to see big city lights!'

En route to Lanchow (Lanzhou), the engine gave them continuous trouble. After a couple of days, Needham and Huang managed to get a lift on a northbound army truck, which took them part of the way, and finally they arrived at Lanchow. They made contact with Teichman, took a gasket from him and bought another. There was a pile of telegrams from London, and Needham made haste to get back to Chungking. But he was offloaded several times from the plane at Lanchow airport to make way for consignments of the famous melons of Hami being sent to Generalissimo and Madame Chiang Kai-shek. The last thing he saw of Alley was his broad grin through the open door of the DC4, remarking, 'Don't believe you're really off until you're actually in the air, my boy!'

Still to come for Needham were arduous journeys to the south-east and the south-west in 1944; and a second south-western journey, followed immediately by a northern journey, in autumn 1945. His wife Dorothy arrived in February 1944 and accompanied him on the journeys to the south-west and to the north. She was the Chemical Adviser on the Scientific Staff. His final big journey was to the east in spring 1946, accompanied by Lu Gwei-Djen.

On the morning of 8 April 1944, Needham began a long, fifteen-week journey to the south-east in the S-BSCO van. He enjoyed the exciting passage of the aptly named Seventy-two Bends Highway in the descent of a steep mountain. His aim was to send the van ahead on a flatcar at the railhead at Tushan, and there to board the passenger express train. As Huang noted, 'Needham loves trains, and this was his first train ride in China. He was full of enthusiasm. . . . At day break it went through a series of switchbacks and came down the mountain into a narrow valley. From then on, all day our eyes feasted on a panorama of unending successions of karst peaks [characteristic of limestone regions] which rose like surgarloaf islands from the flat valley floor.'

When, a few days later, they boarded another express going to Hengyang, they met a group of American airmen. With them was an interpreter, Miss Chou Boa-ling, who had a fine soprano voice and a vast repertoire of folk-songs. Needham was so intrigued that he had Huang write down the lyrics and tunes, which they read and translated shortly after. As always, there were holdups, some lasting several days as the van demanded attention. On one occasion, they had to spend three nights in a little inn. It was raining hard. Stuck indoors, Needham taught Huang Gaudeamus Igitur, the Communist Internationale and the Nazi Horst Wessel song.

It was Needham's determination that kept them going. Not only were there the physical difficulties to overcome – 'a tiring day, the gradients so bad, the scenery quite exhausting', and 'Rainy and drizzling all day. Road surface the worst yet met with, traffic quite heavy (too heavy) all running on pinewood gasoline. Just before reaching Chienyang came across a bridge with the boards all quite gone and many rotted to pulp. Had to fix it before crossing' – but also the threat arising from spending time in a disease-ridden country – 'between Yungan [Yongan] and Sanyuen passed many small parties of sick soldiers being driven on by NCOs, themselves hardly able to stand. . . .' Bubonic plague endemic here, of course, also malaria. Many looked at the point of death.' There were the indignities he suffered in some unclean rooms – 'awful night. Mosquitoes exercised unbelievable ingenuity in getting into my net . . . Bugs to be dealt with about 3'. But that was rare.

What was particularly worrying was the possibility of being captured by the Japanese. South-east China formed an enormous salient between Japanese-held territory – north-east China to the north and Indochina to the south. During the middle and latter part of 1944, a successful Japanese offensive sealed off south-east China from Chungking (Chongqing). Aware of this, Needham's aim was to get his colleagues and vehicle back through the Hengyang bottleneck in time. He succeeded – but only just. Two days after they had crossed, the great bridge at Hengyang was blown up.

On 7 June, they read in a newspaper with great excitement that the Second Front in Europe had been opened. On the following day Huang had a shock. Needham informed him that he had been awarded a scholarship for postgraduate study by the British Council. He would have to leave for England by the autumn. Needham asked him to find a replacement. He invited Tshao Thien-Chhin (Cao Tianqin), whom he had met in Chengtu (Chengdu) in February 1943 as a fellow employee of the CIC Technical Institute. They both enjoyed German *lieder*, and Huang admired his 'beautiful' Mandarin accent.

On 25 July, Tshao, who afterwards became head of Academia Sinica's Biochemical Institute in Shanghai, arrived in Chungking as the Needhams were preparing for a journey to the southwest – to Kunming and western Yunnan. These journeys were of critical importance not only for Needham, but also for Chinese science. Huang's perception of what had the greatest impact on Needham was threefold: first, the land and its people. It was not only the amazing beauty and diversity of the country revealed to him in his travels, but also his deliberate contacts with the people – peasants, workers, shopkeepers, soldiers – whose poverty and hard lives, industry and determination, he with his radical beliefs could appreciate.

Second, were the historical links with the past. These were everywhere and enabled him to find an answer to his key question of why it was that modern science did not develop in China. With this was linked, third, the amazing fertility of activity of the scientific and intellectual community.

What had Needham achieved during his three years in China? The details are contained in his reports on the working of the S-BSCB. The only link with the Western world was the air link with India across the 'hump', provided by the US Air Force, and later the RAF. Despite great difficulties, contact was always maintained between scientific and technical institutions and individuals in the different countries, including the exchange of information, specimens, cultures, seeds, etc. This included the arrangements of special research services, for instance, the preparation of thin rock-sections, chemical analyses, map printing. Selected scientific and technological correspondence was transmitted through diplomatic channels.

To ensure an understanding of the needs and working conditions of the scientists and technologists, and of the contributions their research and products could make to the West, the office staff spent a considerable part of time on travel, as did

Needham. In the S-BSCB first year, there were visits to ninety-one institutions: in the second and third years, 205 visits.

Apart from the supply of urgently needed equipment to the research institutions, assistance was given in sending Chinese scientific and technical publications to the West. Original scientific manuscripts were also sent, enabling the Chinese to continue their contribution to world science. Following his travels, Needham gave detailed accounts of what the Chinese researchers were doing in a series of articles published in *Nature*.

As Needham was preparing to return to Europe, he referred to the special problems he had had to face in China,

where the language problem offers special obstacles of its own, where in the war capital housing is an almost insoluble problem and telephone and other communications are very inefficient, where a visit to a factory only a few miles away may involve a couple of days' journey by sedan-chair, and where, owing to the isolation of the country by blockade, all the remaining worn-out motor vehicles are subject to constant time-consuming breakdowns, often in remote places, owing to lack of spare parts – the difficulties of accomplishing anything may be dimly envisaged. However, something has been done, even though it can be regarded only as a beginning.

In his first-year report, Needham had commented briefly on the inevitable growth of organizations for international scientific and technical co-operation. His proposal for an international science co-operation office originated partly from his experiences in China, and partly from the widely held theory of the worldwide commonwealth of science. It led also to the placing of the 'S' for 'Science' in the United Nations Educational, Scientific and Cultural Organization (UNESCO).

In a farewell tribute to Needham, Wang Ching-Hsi (Jingxi), Director of the Institute of Experimental Psychology, Academia Sinica, wrote of his sadness and that of scientific workers at Dr Needham's departure: 'He knows our merits and our defects and tells us of them frankly, except for those things which it is not convenient for a guest to say. His practical assistance to us can never be forgotten by our scientific workers.'

And Fu Ssu-Nien (Sinien), Director of the Institute of History and Philosophy of Academia Sinica, and Acting President of Beijing University, agreed that the failure of modern science to develop in China was due to differences in the social and political structure and environment and not to any inherent ineptitude of the Chinese for science. He praised Dr Needham for his understanding; 'instead of seeing our poverty and simplicity he saw our perseverance; instead of our backwardness, our future hopes.'

In further tribute to him, astronomers at the Purple Mountain Observatory named a 'minor planet' or 'asteroid' they discovered Li Yüehsê (Li Yeuse), after his Chinese name.

# 6

# UNESCO and science co-operation

*The core function of governance is that of community
learning at the world level; of organizing and
facilitating that learning. After all, the learning
capacity of a society determines its ability for
sustained progress and development.*

Knut Hammarskjöld, 1990

A treasured historical statement stands in a corner at the Needham Research Institute in Cambridge. It is a massive desk, ornamented with anchors in the corners. It had been used in Paris by the admiral in change of German naval operations during the Second World War. It stood then in the old Hotel Majestic in the Avenue Kléber, the seat of the occupying Nazi power.[1] The desk, which Needham was to inherit, was in Paris, but he was in London, to which he had travelled in considerable haste from China following an urgent telegram from Julian Huxley dated 5 March 1946. It was an invitation to join the Preparatory Commission of UNESCO with responsibility for building up a Division for the Natural Sciences. He found the new international organization in rather cramped quarters at a house in Belgrave Square, so restricted, he recalls, 'that I remember having to interview potential secretaries in a bathroom'.

London, during the final destructive years of the war, was the exciting, constructive centre of the resistance to Hitler's armies. It housed seven governments-in-exile – Belgium, Czechoslovakia, Greece, Netherlands, Norway, Poland, Yugoslavia, plus General de Gaulle's Free French, and, as early as in 1940, an estimated 200,000 refugees, including diplomats and intellectuals from all disciplines. The government asked the British Council to look after their cultural needs, including education.

---

1.  Needham told this story when he was awarded the UNESCO Einstein Gold Medal in July 1994.

Thus, on 16 November 1942, R. A. Butler, President of the Board of Education of England and Wales, was able to convene the first meeting of the Conference of Allied Ministers of Education (CAME) at Alexandra House, London, which was eventually to foster the birth of UNESCO. However, the agenda made no reference to setting up an international educational organization. But at the second session in January 1943, there arose the need to ensure that the work of the International Institute of Intellectual Cooperation (IIIC) should be given full support after the war.[2]

There was a speedy follow-up. In May 1943, during the CAME fourth plenary session, a fifty-four-page report, 'Education and the United Nations', prepared by the London International Assembly (LIA)[3] and the Council for Education in World in Citizenship (CEWC),[4] made three main recommendations: the creation of a temporary Bureau for Educational Reconstruction; the appointment in Germany of a High Commissioner (to ensure Nazi and military influence were utterly eradicated); and the establishment of a permanent International Organization for Education. This last proposal was regarded as premature, but only two months later CAME itself was being seen as the seed of such an organization.[5]

By the end of 1943, CAME enlarged itself constitutionally by giving full delegate status to those countries until then represented by observers. The United States Government accepted

2.  The IIIC was founded under the auspices of the League of Nations after the First World War, to promote the ideals of international understanding, but it had not been effective.
3.  The LIA arose from activities in 1940 and 1941 of a Continuation Committee to promote an international inter-allied Office of Education after the war.
4.  Created in 1934 to alert 'the rising generation to the threats to world peace'.
5.  The background story of how the only active, successful 'specialized agency' of the League of Nations, the International Labour Organisation (ILO), set up in 1919 after the First World War, came to be regarded as a model body, and details of the role of the CEWC and the LIA in helping in the conception of UNESCO, is told by Derek Palmer in Chapter 3 of *Peace Through Education*, Lewes (United Kingdom), The Falmer Press, 1984.

an invitation to participate fully, stressing that 'the governments collaborating in the Conference should take steps, with other interested governments, to seek the best means for establishing a United Nations agency for educational and cultural reconstruction'. The founding conference of UNESCO took place on 1 November 1945 and a few weeks later, in December 1945, CAME dissolved itself.

Much help was given by the United States representatives, reinforced by the statesman, Joseph Fulbright, who went to the United Kingdom with a draft for a United Nations cultural organization. But CAME had become aware that science was part of culture and in 1943 formed a Science Commission. Sir Henry Dale was Chairman and J. G. Crowther, of the British Council, Secretary. The commission quickly found that it was not possible to receive from the governments in exile any clear and comprehensive statement of what might be their post-war scientific needs. However, with the assistance of Brigadier R. A. Bagnold, geographer and publicist, a plan for 'skeleton inventories' was drawn up. It was highly successful.

Julian Huxley, biologist and scientific humanist, was appointed as first Director-General of UNESCO, persuaded to allow his name to go forward by Ellen Wilkinson, M.P., who had become Minister of Education in the post-war Labour Government. He and Needham were old friends. They shared a similar political outlook and sense of social responsibility as scientists. Huxley had been the first president of the Association of Scientific Workers in 1929, and Needham represented the AScW on the Cambridge Trades Council, until he went to China in 1943.

Needham was the obvious person for Huxley to invite. The nature of his job as Scientific Counsellor in Chungking turned out to be directly connected with UNECO becoming UNESCO, and with what was to happen at UNESCO later.

On 29 December 1943, Needham wrote to Dr T. V. Sung (Song), then China's Foreign Minister:

In order to make more precise our conversation of yesterday, I would like to indicate . . . what I feel is necessary in post-war scientific co-operation. I believe that the time has gone by when enough can be done by scientists working as individuals, or even in groups organized as universities, within individual countries, and contacting each other across national boundaries, as private individuals. Science and Technology are now playing, and will increasingly play, so predominant a part in human civilization, that some means whereby science can effectually transcend national boundaries is urgently necessary.

The Science Co-operation Offices which have already been set up in the capitals of the United Nations are a piece of machinery which ought to continue after the war. The need for contact can be met neither by instituting 'scientific attachés' at all Embassies, for that would be too subject to diplomatic formalities; nor by sending from one country to another industrial scientists whose loyalty is to particular commercial interests, for their advice could hardly be unbiassed or disinterested. What I should like to see would be some kind of World Science Co-operation Service, whose representatives in all lands would have semi-diplomatic status and full governmental facilities in transportation and communication, but who would be drawn from either academic or governmental laboratories, and would therefore be free from commercial entanglements.

One of the immediate aims for such a world service would be the conveyance of the most advanced applied and pure science from the highly industrialized Western countries to the less highly industrialized Eastern ones; but there would be plenty of scope for traffic in the opposite direction too.

Following this, Needham produced three long memoranda urging the formation of an international science co-operation service. He sent these papers to many 'opinion-forming' people throughout the Western world. He was convinced that many scientists felt that it was essential to have an intergovernmental body to assist the organization of the natural sciences in all countries.

The first memorandum was issued from Chungking in July 1944; the second, addressed to the British Parliamentary and Scientific Committee, was written and circulated in London, where he had returned for consultations, in December 1944; and the third from Chungking in April 1945, following a stimulating visit by Needham to the United States. Some 500 copies were duplicated and distributed, and read by such decision-makers as Clement Attlee, Henry Wallace and V. M. Molotov. Needham, invited to the Conference in Moscow on the 200th Anniversary of the Academy of Sciences of the USSR, took copies with him, and presented one to Dr Harlow Shapley, who later became scientific adviser to the American delegation at the founding of UNESCO. Mr Archibald MacLeish, leader of the American delegation, proposed that the word 'science' should be included in the title of the new organization.

This suggestion for the name 'United Nations Educational, Scientific and Cultural Organization' won wide acceptance speedily. In November 1945, the insertion of the 'S' into UNESCO was carried by a large majority at the constitutive General Conference in London.

Meanwhile, Huxley, determined to express his imposing task as Director-General of the new organization, wrote a sixty-page pamphlet entitled, *UNESCO, Its Purpose and Philosophy*. In this,

besides stressing its obvious duties in promoting cultural exchanges and giving help to the educational systems of backward [or, as we say now, 'underdeveloped'] countries, I maintained that it could not rely on religious doctrine – there was strife between different religions and sects – or on any of the conflicting academic systems of philosophy. UNESCO, I wrote, must work in the context of what I called *Scientific Humanism*, based on the scientific facts of biological adaptation and advance, brought about by means of Darwinian selection, continued into the human sphere by psycho-social pressures, and leading to some kind of social advance, even progress, with increased human control and conservation of the environment and of natural forces. So far as UNESCO was concerned, the process should be guided by humanistic ideals of mutual aid, the spread of scientific ideas, and by cultural interchange.[6]

Needham found himself in broad sympathy with that statement, although as a result of opposition to Huxley's document by the historian, Sir Ernest Barker, an ardent churchman, who argued against what he described as an atheistic attitude disguised as humanism, it was never issued as an official document. Later, Huxley accepted that Barker was right. 'Though UNESCO has in fact pursued humanistic aims, it would have been unfortunate to lay down any doctrine as basis for its work.'[7]

Needham was not quite happy with the declaration that opens the Preamble to the Constitution of UNESCO: 'Since wars begin in the minds of men, it is in the minds of men that the defences of peace must be constructed.' He found this otherwise striking statement far too idealistic, since he believed wars were much more likely to begin from economic and social causes. Dr Richard Hoggart, at one time ADG for Social Sciences at UNESCO, has referred to the phrase as having 'all the resounding opacity of such phrases at their most dense'. He states that

6.   Julian Huxley, *Memories II*, p. 15, London, Allen & Unwin, 1973.
7.   Ibid., p. 16.

the words were first drafted by the British journalist, Francis Williams, and spoken at the Founding Conference in 1945 by the British Prime Minister, Clement Attlee, and were later modified by Archibald MacLeish, the American poet and librarian. Hoggart wrote that 'it has haunted, inspired or befuddled UNESCO's Councils ever since'.[8]

Needham found the description of the purpose and functions of the Organization (Article 1) induced a kind of split personality, amounting sometimes to a kind of schizophrenia in the Organization. Thus, in and around the new body there were many who claimed that UNESCO's activities should be dominated by their making an immediate and direct contribution to peace and security. But, scientific and cultural activities could make only a rather long-term and indirect contribution to those objectives. Surely that was the task of the United Nations Security Council? UNESCO's task lay rather in diffusing and advancing knowledge by encouraging international co-operation in all the fields of science and learning.

The implications of this divergence of interpretation caused a division of opinion which expressed itself both in the delegations and the secretariat. For instance, those insisting on the dominance of peace and security believed that UNESCO should not fund international scientific meetings because they did not contribute immediately to peace. The funds should be made available for mass propaganda for peace. Needham was utterly against such 'a complete distortion of the purposes for which UNESCO was originally set up'.

He insisted that there was no field with a stronger tradition of international community work than that of the natural sciences, and that it could be taken as a model for other fields. His scientific programme was built on two pillars: the Field Science Co-operation Offices, and financial aid to the International Council of Scientific Unions (ICSU) and the individual scientific unions represented on it.[9]

8.   Richard Hoggart, *An Idea and Its Servants: UNESCO from Within*, p. 27, London Chatto & Windus, 1978.
9.   ICSU is a non-governmental organization founded in 1931 to bring together natural scientists in international scientific endeavour. In December 1994 it had a national membership of ninety-two multidisciplinary bodies (scientific academic or research councils)

Obviously, his Chinese experience in creating the Sino-British Science Co-operation Office was the model for the former. He wanted to place science co-operation offices in strategic centres in Third World countries, able to provide speedy help to hard-pushed scientists and technologists isolated from the main centres of science and technology by, for instance, providing them with needed apparatus, literature or specimens, putting them in touch with colleagues, and seeking quick publication of research in key scientific journals. Offices were set up in such places as New Delhi, Shanghai, Montevideo, Cairo, Istanbul, Manila, Bangkok and Jakarta. As a director of each office, he sought recruits from around the world. The person on his staff at UNESCO responsible for the whole operation was the much experienced Dr Lu Gwei-Djen.

UNESCO was required to seek help in technical matters from non-governmental organizations (NGOs), international agencies concerned with subjects within its terms of reference, and, where necessary, to help in the creation of new ones. The first NGO to be attached to UNESCO was ICSU, which was provided with a liaison office at UNESCO Headquarters and with substantial grants in aid to the affiliated unions. Needham recalled that those were the days when an enormous amount could be effected by telephone. For example, when he learned that the sea-water pumps at the Stazione Zoologica at Naples were almost worn out, he rang the British Minister, Sir Stafford Cripps, whom he knew, to point out that many such pumps, owned by the American army, were rusting on the quay beside the Institute. Within a week, they were being installed.

Again, at a conference in Poland, at an opening reception a

---

and twenty-three international, single-discipline Scientific Unions. It has also twenty-nine Scientific Associates. Financial resources consist of Members' contributions, a UNESCO subvention, grants and contracts from other United Nations bodies, foundations and agencies. Its total budget is over $15 million a year. ICSU acts as a focus for exchange of ideas and information and development of standards. Hundreds of congresses, symposia and other scientific meetings are organized each year worldwide. Publications include a wide range of newsletters, handbooks and journals. The Executive Director is Julia Marton-Lefèvre. The international secretariat is at 51 boulevard de Montmorency, 75016 Paris, France.

complete stranger embraced him and shouted, 'Behold the saviour of Polish mathematics!' Astonished, he was reminded that the Nazis had left Poland without any machines capable of printing mathematical formulae, and that a phone call of his from UNESCO had been responsible for providing them.

Needham's division was active, also, in the applied sciences: for instance, collaborating closely with the World Health Organization (WHO) and with the Food and Agricultural Organization (FAO). Plans in hand, as his term at UNESCO was ending, were for a conference of all international NGOs in the engineering sciences to work out something parallel to the federation of unions in the pure sciences. Work had begun, also, on the popularization and social implications of science.

During his two years at UNESCO, Needham was much concerned with internal problems, matters which have plagued the organization throughout its existence and continue still. For instance, the serious tendency for administrative management to act as policy programme-forming sections rather than as programme-facilitating and service sections. When taking up this and other matters with Huxley, he warned that unless these were dealt with wisely the Director-General was increasing his 'dangerous position of isolation'.

The matter he took very seriously was what he regarded as the challenge to his own personal position and reputation. On the evening of Friday, 2 January 1948, he had what he described as 'a very pleasant talk' with Huxley. Two days later, he wrote him a detailed letter because 'events were hurrying rather fast'. Needham urged Huxley to make his reply 'as frank as you like, knowing that nothing could spoil our deeply cemented friendship and my loyalty to you and your plus fourteen years'.

Needham was greatly distressed to learn that the highly personal letter he had himself taken to the Hotel Bristol, where Huxley was temporarily in residence, seemed to have been lost by Huxley before reading it. Hurriedly, he produced a copy and begged Huxley not to leave it lying around. The letter concerned Needham's status. He had for long felt deeply that the UNESCO directorate policy-making level was not strong enough and that the meetings of the heads of sections were far 'too large, amorphous and weak'.

Under consideration was the proposal to appoint an Assistant Director-General (ADG) for each of the main activities of

UNESCO. For Education and Culture, Needham saw no difficulty. But he felt that the 'S' for 'science' stood also for 'snag',
because it was difficult to combine both the natural and the social sciences under any common numerator.

He believed that the French cosmic-ray physicist, Pierre
Auger, on the Executive Board, had wide enough knowledge
and interests to act as an overall ADG. But where did he himself
come in? He was the Head of the Natural Sciences section and
due back at Cambridge in April. He believed that the decision
to return could not be reversed, except by the British Government. Meanwhile, given 'the patronizing and hostile attitude of
our Inhibitor-General',[10] for whose intellectual attainments he
had scant regard, what Needham wanted was to be given more
'face'. He explained that the Natural Sciences section was running well, and if he were to be named as ADG for Science, it
would not be necessary to appoint a new Head of Section. He
could handle that together with a general responsibility for the
Social Sciences.

He stated that he would 'feel unhappy and distressed because it would reflect publicly' on him if an ADG post were to
be created immediately upon his departure. 'I shall be put down
as an interim stopgap unworthy of larger responsibility. I hinted
at this at dinner on Friday evening, but I don't think you understood what I was getting at. I don't feel I deserve this undesir-
able reflection.'

However, he went on, if the decision was definitely to have
an ADG for Science, his uneasiness could be met by appointing
him for a token period of a few weeks, after which he would go.
Despite this urgent plea, the final decision about the appointment of ADGs was not taken for some time after he had gone.

When, on 20 April 1948, Needham left UNESCO to return
to Cambridge, his staff presented him with the German admiral's desk at which he had worked at the Hotel Majestic, the first
official UNESCO headquarters. His links with UNESCO remained. In 1949, the Executive Board conferred on him the title
of Honorary Counsellor, and in July 1994, the Director-General of UNESCO, Federico Mayor, presented him with the
UNESCO Einstein Gold Medal during a special ceremony at
the Needham Research Institute in Cambridge.

10. I have been unable to establish the identity of this individual.

# 7

# 'Peasants' and Master

*Brethren, behold how sweet and pleasant a thing it is, to dwell together in unity.*

The Psalmist

Back in academic Cambridge in 1948, Needham was once again in his college room, K1, but lived in a house at No. 1 Owlstone Road into which he crammed as many books, papers, and objects as he could. He found the college atmosphere had changed. He recalled that after he was elected a Fellow his life had revolved much more about the Biochemical Lab, and in it, than it did in the college. He had his lunch and tea in the lab every day and got on well with some brilliant colleagues. It was there that he had met his first wife, Dorothy.

He felt the change he was experiencing was of a world becoming much more open. There was a great difference in the atmosphere of the Senior Combination Room and conversations in Hall. In prewar days he had been stifled by the conventionality of conversation. Any expression of what was regarded as 'subversive' opinion was impossible, as was any objective discussion of political and social problems. That had changed: conversation was freer, albeit conservative opinions were still dominant.

There was, also, a battle for more openness in running the college. Needham could not forget that for many years after he became a Fellow he found almost all the older Fellows impossible to talk to. He was very isolated and, as he put it, 'was altogether an Ishmael in College society'. He was an 'absolute outsider', and was deliberately cold-shouldered. And not only in College.

In 1934, he had a visit from W. F. Tisdale, representing the Rockefeller Foundation, which had initiated a policy of biological progress, conceived around the idea of technology transfer from the physical sciences to biology. As a result, in the following year, Needham together with J. D. Bernal and Dorothy Wrinch proposed the setting up of a research institute bridging the disciplines of physics, chemistry and biology. In May 1935, Tisdale again visited Cambridge, accompanied by Warren Weaver, responsible for Rockefeller's natural sciences programme. Weaver believed strongly that the future of physical science lay in its application to biology, and in 1938 he coined the term 'molecular biology'.

However, during his 1935 visit to Cambridge, although he found no doubts that Needham and Waddington were well qualified to run the proposed institute, he was disturbed by the 'English attitude to Needham'. Both Sir Henry Dale, secretary of the Royal Society, and Sir Edward Mellanby, secretary of the Medical Research Council, key leaders of the scientific establishment, were against support for Needham on socio-political grounds.[1]

But 'a revolt' was to take place at Caius. Not arising from a coterie of 'Ishmaels', but from a legitimate desire of many Fellows to have a say in the administrative workings of the College. The College Council had what was essentially a self-perpetuating membership. It consisted of all the Tutors, who were treated as permanent officers, in addition to the Dean and the Steward. The Council elections were not contested. However, in October 1950, a 'Peasants' Revolt' took place at the General Meeting of the College. It resulted in two senior Fellows being replaced by two younger Fellows – Peter Bauer and Michael Swann. Both were later to become Peers of the realm. *'Bauer'* is German for 'peasant', hence the name of the 'revolt'.

The original Peasants were eight in all – four were very senior Fellows, and without their support and that of some other senior Fellows they would not have been able to gather seventeen votes as they did. Needham was, of course, a Peasant. In a sense, the affair reminded him of his cautionary letter to Huxley at UNESCO, when he pointed out the inherent conflict between the administrative and the academic approach to behaviour.

1. 'The Discourse of Physical Power and Biological Knowledge in the 1930s: A Reappraisal of the Rockefeller Foundation's "Policy" in Molecular Biology', *Social Studies of Science*, Vol. 12, 1982, pp. 341–82.

Interestingly, such a 'revolt' did not occur in any other Cambridge College. But it did expose the manipulating behaviour of the Caius 'Old Guard'. As Christopher Brooke makes clear, 'the Peasants' Revolt came in on a tide of genuine academic idealism which was to generate an admirable enthusiasm for lofty aims; it generated, also hard feelings, personal misunderstandings, misery and faction which were only finally laid to rest under the Mastership of Needham'.[2]

The objectives of the 'Peasants' Revolt' were to secure some rotation both of Council membership and of Tutorships, and sufficient recognition by the Council of 'the work or the needs of Fellows who served the statutory purposes of the College by research rather than by administration'.[3]

Recognition that the College was not only a place of education but also a centre of learning and research was very largely secured within a year or two. These important changes in college life were to allow Needham to become President in 1959 and Master in 1966. These high offices reflected the significant regard the College had for him as a scholar, eminent in his field.

The office of President is a curious one. Originally, he was the President of the Fellows, their representative in relation to the Master, but when statutes were drawn up in the 1920s, the President in effect became the Vice-Master. During the Master's absence, he guides college affairs: he presides at dinner in Hall and dessert in the Combination Room, and fulfils many social and ceremonial duties.

Needham followed the eminent solid-state physicist, Sir Nevill Mott – later, in 1977, a Nobel laureate – as Master. Mott never gave up his scientific work: he remained Cavendish Professor throughout his term of office, from which he resigned in 1961. There was no real opposition to Needham's election as Master. It is very unusual for a President to be elected Master, and he was one of the very few to be so. At an adjourned meeting of Fellows on 3 December 1965, for the purpose of pre-election of a Master, by a majority of votes of the forty-five Fellows present 'Noel Joseph Terence Montgomery Needham, Sc.D, F.R.S., Sir William Dunn Reader in Biochemistry, was pre-

2. Christopher Brooke, *A History of Gonville and Caius College*, p. 272, Woodbridge (United Kingdom), Boydell Press, 1985.
3. *The Caian*, November 1985, pp. 37 et seq.

elected to the Mastership as from 1 February 1966 until the end of the academic year following the expiration of seven years from 1 February 1966'. On that day, in the College chapel immediately after divine service he made the required statutory declaration and became Master.

As Master, Needham made an impact quite different from what was expected of 'so radical' a person. He recognized very early that the Mastership is very much what the person makes of it. As he wrote,

If you want to play a big part in administration, sit on all the Committees there are, and invent a lot of new ones, then you are quite free to do so; if on the other hand you have something better to do, you may find that everything works all the better the less interference you apply to it. This was my principle, following the Taoist conception of *wu wei* (no action contrary to nature); and I left everyone to do their jobs – Bursars, Tutors, Deans, etc. – which they did extremely well without any fussing from me.[4]

He intervened only on the rare occasions when something strange or unusual occurred. He likened the position of the head of a Cambridge college

to someone sitting in front of an enormous console of some macro-computer, where there were hundreds of buttons to press and hundreds of coloured lights to come on and off. Most of the time the lights showed green, and there was no need whatever to do anything, but every now and then a red light would come up, and you would know that some bearings were running hot somewhere, so that the question was, what exactly was the best thing to do. That was where judgement came in, and the knowledge of whom one could trust in particular things. It soon became very evident that it was no use trying to change the character of any of our 75 Fellows, the Governing Body of the College. One had to accept each one as Nature had made him, and do all one could to utilise the best aspects of each particular person for the general good.

He had used a similar metaphor during his early days in Kunming.

When Needham was reading seventeenth- and eighteenth-century Latin works on medicine, he was 'much impressed by the phrase *opinio conciliatrix,* said of some proposal which might

4.   *Festschrift Autobiographical,* pp. 45 et seq.

bring about agreement between those who held entirely, or what seemed to be entirely, contradictory theories about the problem in question'.[5]

That kind of effort was his aim as Master. He sought to ensure that there was a good atmosphere not only among the Fellows, but also among the students and the staff. He introduced what he described as some 'new "ancient" customs'. For instance, he instituted a meeting each term and lunch in the Lodge of the College Officers and the Heads of staff departments, so that their relationships became much closer and more cordial. Again, he set up the Domestic Management Committee to meet the desire of undergraduates for more share in the management of the College. There were two or three tea-time meetings each term at which the students' elected representatives joined the College Officers and the Heads of staff departments appropriate for the subject under discussion. Needham found this 'an extraordinary good thing because you got a cross-section of attitudes throughout the College'.

As they were the Hall of the Annunciation of the BVM, commonly known as Gonville and Caius College, he decided to have a special evensong on the Feast of the Annunciation and invited the boys of the Perse School – founded by their great medical Fellow, Stephen Perse, in 1615 – to sing, followed by a party in the Lodge.

He regretted one 'non-change' – the admission of women as students and Fellows. However, he believed that time would come. And so it did, when Dorothy Needham became the first woman Fellow in October 1979.

He was reminded by the editor of *The Caian* that he had written that 'there is something divine about a committee'. Was that what he had found in practice? He replied that there was a great deal of truth in that. He remembered that Conrad Noel of Thaxted in his sermon on Trinity Sunday always used to say, 'You see, the universe itself is governed by a committee.'[6]

Needham believed very strongly in collective wisdom, but he sometimes thought 'there is something wrong with our way of conducting meetings . . . people's personalities seem to undergo a change for the worse when they sit round a table in

5. *The Caian*, November 1976, p. 34.
6. 'Interview with Joseph Needham', *The Caian*, p. 37.

formal meeting. We tend to become formalistic, stilted, rhetorical or unduly emotional, short-tempered and easily offended, sometimes disingenuous, even insincere, rarely "thinking aloud" in a relaxed and objective way.' He had no answer to this, but just tried to make the best of it.

Meanwhile, with his Masterhip his financial situation had become easier. His annual stipend was to be £4,150 while he held no other paid office. It was agreed, also, that he should retain his rooms on K staircase. As a quid pro quo, his offer of the top floor of the Lodge, excluding the Gonville Room, for conversion for resident and non-resident Fellows, and his ex-gratia offer of a room for a research student were accepted.

On 13 March 1966, instead of Evensong, a Roman Catholic mass was held in the chapel at 6.15 p.m. It was the first such mass since the eighteenth century.

He served as Master for ten years, then still as a Fellow, he kept Room K2 in Caius Court. He had enjoyed his time. Especially typical was the occasion in the summer of 1976, when at a party at the lodge, he and his guests danced Scottish reels in the garden. Those dancing included Dorothy, Gwei-Djen, Hashimoto Keizo, a shy young Japanese, Nathan and Carole Sivin, Francesca Bray and her husband, Peter Burbridge and his wife Muriel, Scottish and gifted with the second sight so she always knew when sailors were not coming back.[7]

Needham was pleased to give up as Master, because he found it imposed a physical strain. He had succeeded in keeping the College 'a happy ship', at a time when the number of Fellows increased from 61 to 77, there were 500 students and about 200 technical staff, without taking into account the Old Caians, 'who compass about like a great cloud of witnesses'.

Needham became 'a symbol of the College and its meaning for a whole generation', commented Christopher Brooke. 'Few of our Fellows have been so deeply read in the history of the College or so devoted to aspects of its traditions as Needham, the devout Caian and romantic Anglo-Catholic.'[8]

7.   Joseph Needham, letter to Horace Judson, 12 December 1991.
8.   Brooke, op. cit., p. 277.

# *S C C :* the rivers pay court to the sea[1]

*The love of anything is the fruit of our knowledge,
and grows as our knowledge becomes more certain.*

Leonardo da Vinci

At long last more at ease financially, Needham could devote himself to his systematic treatment of the Chinese contributions to the sciences, technologies and medicine in ancient and medieval times. His aim was to cover the entire pre-seventeenth-century world. His cut-off period was the beginning of what he came to term as 'modern science'.

The project, *Science and Civilisation in China (SCC)*, had been conceived originally around 1939 in conjunction with Lu Gwei-Djen, who became his chief collaborator in 1957 and remained his right hand until her death in 1991.

Through his special relationship with her and the two other Chinese graduate students, he first recognized that the civilization of China was just as subtle, complicated and ancient as that of India or the Western world. They caused him to ask what is referred to as the 'Needham question'. Why did modern science originate only in Europe? His answer to this was to consume his mental and physical energies for the remainder of his life. He would often rephrase the question by asking why modern science did not originate in China, given their significant technological inventions. He went on, also, to ask whether the science and technology of the past had contributed in a direct genetical succession to that movement in seventeenth-century Europe from which modern science emerged.

1. 'Image of the ancient and medieval sciences of all the peoples and cultures as rivers flowing into the ocean of modern science', old Chinese saying.

In more detail, Needham's own words were:

What exactly did the Chinese contribute, in the various historical periods, ancient and medieval, to the development of Science, Scientific Thought and Technology? The question can still be asked for later periods, though after the coming of the Jesuits to Peking in the early seventeenth century, Chinese science gradually fused into the universality of modern science. Why should the science of China have remained, broadly speaking, on a level continuously empirical, and restricted to theories of primitive or medieval type?

How, if this was so, did the Chinese succeed in forestalling in many important matters the scientific and technical discoveries of the *dramatis personae* of the celebrated 'Greek miracle', in keeping pace with the Arabs (who had all the treasures of the ancient western world at their disposal), and in maintaining, between the third and thirteenth centuries, a level of scientific knowledge unapproached in the west? How could it have been that the weakness of China in theory and geometrical systematization did not prevent the emergence of technological discoveries and inventions often far in advance (as we shall have little difficulty in showing) of contemporary Europe, especially up to the fifteenth century?

What were the inhibiting factors in Chinese civilization which prevented a rise in modern science in Asia analogous to that which took place in Europe from the sixteenth century onwards, and which proved one of the basic factors in the moulding of modern world order? What, on the other hand, were the factors in Chinese society which were more favourable to the application of science in early times than Hellenistic or European medieval society?

Lastly, how was it that Chinese backwardness in scientific theory co-existed with the growth of an organic philosophy of Nature, interpreted in many differing forms by different schools, but closely resembling that which modern science has been forced to adopt after three centuries of mechanical materialism? These are some of the questions which the present work attempts to discuss.[2]

Needham's approach as expressed in *SCC* provides one of the most original works of any age. I regard it, also, as his contribution to his autobiography, his reconnaissance of love of China, reinforcing his long-held moral and political beliefs, fixed deeply during the first half of this century, which is his century.

Only the polymath Needham could have undertaken this monumental work, 'formidable to produce and formidable to

2.   Joseph Needham, Preface, *SCC*, Vol. I, p. 2–3.

read', as the historian Lynn White Jr wrote.[3] And Brian Harland, Needham's collaborator at Caius, comments that Needham was fortunate as he was not involved in playing any active part in the faculties of history or oriental studies; thus he was able to avoid the jealousies that seem inescapable from university life and to get on with his work. With no formal training as a historian, he came to be recognized as one of the university's greatest historians.[4]

Needham succeeded in transforming the historian's basic ideas about the definition of science in different cultural backgrounds. He became, beyond challenge, to quote Lynn White Jr again, the world's greatest scholar in the comparative study of civilizations.[5] In one hand he held the Old World, and in the other the New. He sensed that he alone was destined to be the 'bridge-builder', to link them together. Eurocentrism was out and other traditions were no longer to be treated as 'backward', awaiting only the embrace of advanced Europe. For him,

the die is now cast, the world is one. The citizen of the world has to live with his fellow citizens. . . . He can only give them the understanding and appreciation which they deserve if he knows the achievements of the sages and the precursors of their culture as well as of his own. We are living in the dawn of a new universalism, which . . . will unite the working peoples of all races in a community both catholic and co-operative. . . . The mortar of this edifice is mutual comprehension.[6]

Particular aspects of Chinese civilization, such as literature and art, and even economics, had been studied, but science and technology had been ignored. That awaited Needham, and I use 'awaited' in an almost religious sense. He was specially qualified. He knew the language. He had the scientific qualifications to help him evaluate the Chinese contributions. There was no one else anywhere who was capable of doing that. He was familiar, too, with the history of science in Europe, and its social and economic background. Finally, he had an intimate personal experience of life in China, which enabled him to make contact

3.   *ISIS*, Vol. 75, No. 1, 1984, p. 276.
4.   'Joseph Needham: A Personal Impression', *Interdisciplinary Science Reviews*, Vol. 15, No. 4, 1990.
5.   *ISIS*, op. cit.
6.   Preface, *SCC*, Vol. I, p. 6.

with, and to obtain guidance from, a wide range of Chinese scientists and scholars. He pushed himself hard 'fearing that it might well be some time before the same collection of circumstances recurs in another person'.[7] He was prepared to sacrifice much to accomplish his unique project.

His book is the consequence of a rich trawl of the vast and scattered literature that existed and which had never before been digested and built into a single work. In addition to the Japanese and Western literature, there was 'the veritable ocean of the extant original Chinese books themselves'.[8]

With his first collaborator, Wang Ling, whom he had met in 1943 as a junior Fellow of the Academia Sinica Institute of History and Philology, which had been evacuated to Lichiang (Lijiang) in Szechuan, he began in 1948 to lay down the course for the project. They thought seven volumes would be sufficient to cover all the sciences and technologies, but it became very clear that the richness and originality of the material would make that impossible. Each volume would have to be divided into several parts, to be published as specific contributions in themselves. Thus it was that those associated with the project came to speak of them as 'heavenly' volumes and 'earthly' volumes, the former being the original seven, and the latter the separate parts.

About 1970, Lu Gwei-Djen and Needham were obliged to make a major decision: whether to go on 'toiling away single-handed, as it were, producing each volume, in which case the series had no hope of being finished in our lifetime; or whether to associate with ourselves a body of expert and valued collaborators, in which case serious problems of organization would necessarily arise'.[9]

A list of what has been published to date follows:

7.  Preface, *SCC*, Vol. I, p. 6.
8.  Ibid.
9.  Joseph Needham: *The Future of the EAHSL & I – A Personal Statement*, July 1985 (private MS).

*Science and Civilization in China* (SCC), published by Cambridge University Press

| Volume | Title | Author/collaborators | Publication date | ISBN |
|---|---|---|---|---|
| I | *Introductory Orientations* | By Joseph Needham, FRS, with the research assistance of Wang Ling | 1956 | 0 521 05799 X |
| II | *History of Scientific Thought* | By Joseph Needham, FRS, with the research assistance of Wang Ling | 1956 | 0 521 05800 7 |
| III | *Mathematics and the Sciences of the Heavens and the Earth* | By Joseph Needham, FRS, with the collaboration of Wang Ling | 1959 | 0 521 05801 5 |
| IV | *Physics and Physical Technology* Part 1 *Physics* | By Joseph Needham, FRS, with the collaboration of Wang Ling and the special co-operation of Kenneth Girdwood Robinson | 1962 | 0 521 05802 3 |
| | Part 2 *Mechanical Engineering* | By Joseph Needham, FRS, with the collaboration of Wang Ling | 1965 | 0 521 05803 1 |

| Volume | Title | Author/collaborators | Publication date | ISBN |
|---|---|---|---|---|
| | Part 3<br>*Civil Engineering and Nautics* | By Joseph Needham, FRS,<br>with the collaboration of Wang Ling<br>and Lu Gwei-Djen | 1971 | 0 521 07060 0 |
| V | *Chemistry and Chemical Technology* | | | |
| | Part 1<br>*Paper and Printing* | By Tsien Tsuen-Hsuin | 1985 | 0 521 08690 6 |
| | Part 2<br>*Spagyrical Discovery and Invention: Magisteries of Gold and Immortality* | By Joseph Needham, FRS,<br>with the collaboration of Lu Gwei-Djen | 1974 | 0 521 08751 3 |
| | Part 3<br>*Spagyrical Discovery and Invention: Historical Survey from Cinnabar Elixirs to Synthetic Insulin* | By Joseph Needham, FRS,<br>with the collaboration of Ho Ping-Yü<br>(Ho Peng Yok) and Lu Gwei-Djen | 1976 | 0 521 21028 3 |
| | Part 4<br>*Spagyrical Discovery and Invention: Apparatus, Theories and Gifts* | By Joseph Needham, FRS,<br>with the collaboration of Ho Ping-Yü<br>and Lu Gwei-Djen<br>and the contribution by Nathan Sivin | 1980 | 0 521 08573 X |
| | Part 5<br>*Spagyrical Discovery and Invention: Physiological Alchemy* | By Joseph Needham, FRS, and Lu Gwei-Djen | 1983 | 0 521 08574 8 |

| | | | |
|---|---|---|---|
| Part 6 *Military Technology: Missiles and Sieges* | By Joseph Needham, FRS, in collaboration with Robin Yates and other scholars | 1995 | |
| Part 7 *Military Technology: The Gunpowder Epic* | By Joseph Needham, FRS, with the collaboration of Ho Ping-Yü, Lu Gwei-Djen and Wang Ling | 1987 | 0 521 30358 3 |
| VI *Biology and Biological Technology* | | | |
| Part 1 *Botany* | By Joseph Needham, FRS, with the collaboration of Lu Gwei-Djen and the special contribution by Huang Hsing-Tsung | 1986 | 0 521 08731 7 |
| Part 2 *Agriculture* | By Francesca Bray | 1988 | 0 521 25076 5 |
| VII Key parts and publication dates have still to be determined. | | | |

The manner of working for Wang Ling was to seek out relevant texts, prepare most of the draft translations, and be involved in all aspects of the volumes, including research, listing and indexing. There were many detailed joint discussions often resulting in changes before the definitive version was agreed. The ultimate authority was Needham, whose genius lay in the formulation of the problems that would arise in providing his answer to his basic question, and in his approach to their solution. In doing so, he changed the frontier of knowledge as far as the history of science is concerned and defined a new legitimacy for research. The history of science was no longer to be seen as a discontinuous series of individual events. Scientific achievement was a national expression in a national culture, shaped overall by historical circumstances – geographical, moral and political.

Needham's influence lay in promoting the comparative history of science and technology. But for this to make sense, national histories of science and technology were required. As he put it, 'No ancient or medieval science and technology can be separated from its ethnic stamp, and though that of the post-Renaissance period is truly universal, it is no better understandable historically without a knowledge of the milieu in which it came to birth.'[10] Linked with this was public policy, in ancient and medieval China determined by the Emperor and his examination-selected bureaucracy.

From experience, Needham knew that *SCC* would be an impossible task if he had to depend upon translations alone. He recognized that in Chinese, which is a highly isolating and non-agglutinative language, therefore very difficult to understand for a writer in an Indo-European language, the parts of speech are not rigidly differentiated, and a particular word can function as several different parts of speech, depending to some extent on the order of the words in the sentence.

He quoted the sinologist, Friedrich Hirth:

Generally speaking, anyone can translate a chapter of Livy with a grammar and dictionary, but you cannot do that with a Chinese text from Antiquity to the Middle Ages, because there is so much more than the mere meaning of the words and sentences. The European reader must understand, be familiar with, and know the places, the people and the

10. *SCC*, Vol. IV, Part 2, 1961, p. xlvi.

things; he must not only translate, he must identify. Only when he has realized what the author is really talking about, can his translation have the breath of life. Even those who know the language extremely well must also be collectors or, as we might say, students of things, if the things are going to be talked about.[11]

Needham knew that a major part of the necessary resources had never been translated into any Occidental language, and many of the most reliable sinologists had been very careless in their use of scientific and technical terms, even when they understood them. For example, in the *Lun Hêng (Heng) (Discourses Weighed in the Balance)*, written about A.D. 83, in which there is a mention of wine, the translation reads: 'From cooked grain wine is distilled.' But the term in the text is *'miang'*, which means fermented not distilled. The same faults, Needham insisted, applied to many parallel passages.

He found, in addition, that translators had read distinctive Western concepts into Chinese texts in connection with, for example, atoms and the laws of nature. He stressed that certain philosophical terms, such as *Tao (Dao)* (the order of nature) should be repeated in transliterated form and not translated after explanation or definition, when first used. Because Chinese is very rich in homophones, he includes Chinese characters very deliberately in his texts. He insists that a sight of the character intended by the transliteration or romanization is essential for anybody with a knowledge of the language.

But Derk Bodde, Professor Emeritus of Chinese Studies at the University of Pennsylvania, and a collaborator of Needham's, regarded as eccentric the way in which he handled romanization.

The romanization system long used in the English-speaking world for Chinese names and terms is that of Wade-Giles. In this system an apostrophe (') is used to indicate the aspiration of certain syllables. Joseph Needham, however, while following Wade-Giles in the main, replaces the apostrophe by the letter *h*. Thus *t'ang* becomes *thang, p'u* becomes *phu*, etc. The reason for this is not primarily linguistic, as one would suppose. Rather, as Joseph Needham told me, it is that the substitution of *h* for (') saves the trouble of lifting the carriage of the typewriter when typing (') and thus economizes human energy.[12]

11.  Joseph Needham, *The Translation of Old Chinese Scientific and Technical Texts*, No. 1, March 1958 (special issue of *Translation in Asia*).
12.  Personal communication, dated 9 March 1994.

Who was Needham writing for? Interestingly, he did not have sinologists primarily in mind, but all educated people, whether scientists or not, 'interested in the history of science, scientific thought and technology, in relation to the general history of civilization and especially the comparative development of Asia and Europe'.[13]

His work, he believed, made it clear that there was a great Chinese contribution to our understanding of nature. In addition, it made it more clear that Europeans, such as Galileo, did not possess any intrinsic superiority, but benefited from environmental factors, which did not, and could not, operate in other civilizations with a different cultural setting and the different social evolution that implied.

Needham's conviction was that the Chinese were able to speculate about nature at least as well as the Greeks in their earlier period. Then why did not China produce an Aristotle? He argues that it was because the inhibitory factors preventing the rise of modern science and technology in China had begun to operate before the time at which an Aristotle could have been produced.

The American sinologist, Nathan Sivin, a close collaborator of Needham's, points out that Needham 'has given prominence to anticipations of modern science, and drawn upon it for his rubrics, because his aim is to demonstrate the worldwide character of the development of science and technology. But he has consistently probed deeper, and has made sense of many Chinese concepts and shown their connections for the first time.'[14]

*ISIS*, the distinguished journal of the history of science, in its George Sarton Centennial Issue in 1984, devoted a review symposium to the work of Joseph Needham. Lynn White Jr, one of two reviewers, sent an offprint with an inscription to Needham, 'who will recognize that my debt and admiration far outweigh my disagreement'.

His pointed comment was that there were times when Needham seemed to be doing personal penance for the Opium Wars. He went on, 'One respects the exaggerations of a lover. For some forty-five years Joseph Needham has been in intellec-

---

13. *SCC*, Vol. I.
14. Preface to *Chinese Science*, Ed. Nakayama & Sivin.

tual love with China and the Chinese.' However, he remonstrates, the fact that the inhabitants of the Middle Kingdom have traditionally been as contemptuous of all outer barbarians as were the Periclean Greeks does not justify the ethnocentrism of the West during the Age of Imperialism that crumbled only after the Second World War. Needham's work is changing the traditional Western attitude to China, and may do the same as far as Chinese attitudes to the rest of the world are concerned.

Lynn White criticizes Needham's approach based on 'the Marxist belief that all cultural changes, including developments in technology and science, are caused by alterations in social structures'. He expresses dissatisfaction with this aspect of Needham's thesis in his *ISIS* review.

The second reviewer, Jonathan D. Spence, George Burton Adams Professor of History at Yale, takes issue with Needham who insists on the importance of recognizing 'that seventeenth-century Europe did not give rise to essentially "European" or "Western" science, but to universally valid world science, that is to say, "modern" science as opposed to the ancient and medieval sciences.' Spence regards this concept of a branch of science 'getting through' as elusive. However, he declares that Needham's work 'opens up vistas for Ch'ing study few of us had thought of a decade or so ago'. He goes on, 'Again and again, reading his work, my eyes would be drawn to the Chinese characters that accompanied his translations and explanations for the myriad technical and semitechnical terms in which the texts of the tradition are couched.'

He ends with a tribute, a dream,

that perhaps when Needham's great labor was done, Cambridge University Press would find workers to collate from this sweep of volumes an index-glossary with explanations and cross-references, so that all those trying to embark on similar studies would be able to learn systematically what Needham had learned before them. Thus, perhaps in the 1990's, one would begin to be able to make a truer assessment of whether or not China had indeed entered that dialogue of world science, and if so, when. It would be Needham's finest final gift to the spirit of rational enquiry.

Each book is in itself a 'good read'. The author's note that accompanies each volume is a personal confession of great value. And the many half-titles and footnotes, wittily tumbling one after the other, are highly pertinent companions in an intellectual journey filled with surprise and excitement.

Each volume has its dedication; that for Volume I to Lu Gwei-Djen's father reads:

To Lu Shih-Kuo, Merchant Apothecary in the City of Nanking, the first volume is respectfully and affectionately dedicated.

Volume V, Part 2, is dedicated:

To two comrades-in-arms in an age-long struggle,
The use of natural knowledge for peace and love,
Not in the service of hatred and war.
Thang Phei-Sung, Professor of Plant Biochemistry at Chinghua University, Peking, author of *Green Thraldom*, proponent of food for the world – remembering the war-time laboratory among the hills of Tapuchi – and J. DESMOND BERNAL, sometime professor of Crystallography at Birkbeck College, London, author of *Science in History* and the *Social Function of Science*.
Of Loyolan subtlety in Ireland and bred
Three enemies of man he re-interpreted;
Saw world, flesh, devil, black-rob'd walk their rounds
And love's two friends advance a banner red.

Volume V, Part 6, is dedicated:

To the memory of the late, Chou Ên-Lai (1898–1976), leader of the uprising at Nanchong (1927), later Premier of the Chinese People's Republic (1949–1979), constant encourager of this project.

Under the half-titles of the volumes are relevant quotations. Thus, in Volume V, Part 2, appear, in the original language and in English translation, the following, from Friedrich Nietzsche, *Die fröhliche Wissenschaft*, IV, 1886:

Do you believe that the sciences would ever have arisen and become great if there had not beforehand been magicians, alchemists, astrologers and wizards, who thirsted and hungered after abscondite and forbidden powers?

From Michel Eugène Chevreul (1786 to 1889) reviewing Reinaud & Favé in *Journal des Savants*, 1847, p. 219.

Having long been occupied with the history of chemistry, we can clearly see today what difficulties lie in the path of anybody who undertakes to write it. A deep knowledge of the science itself will not suffice unless he has recourse to the ancient, and to the oriental, literature.

From Veritable saying (*hadith*) of the Prophet Muhammad (al-Suhrawardy, No. 273).

Seek for knowledge, even though it be away as far as China.

The relevant quotation for Vol V, Part 6, reads:

Arms are instruments of evil and not the tools of lordship. Only when forced to do so will the good ruler bear them, for he sets peace above all else. Even if he should conquer he will not rejoice, for to do so would be to take pleasure in the slaughter of men. He who takes such pleasure will never accomplish his will in the world.

Truly, weapons are ill-omened things. All beings for ever loathe them. He who has the Tao has no concern with them.

Lords in their halls value the left as the auspicious place of honour; the right is for occasions of misfortune and war. Commanding generals are placed to the right; their adjutants take to the left – for their placing follows the rites of mourning. The slaying of multitudes is matter of grief and tears. Mourning rites therefore greet even victorious commanders.[15]

Mansel Davies has made a tabulation of the contents of the first three volumes to convey some essentials of the range and depth of coverage:

| Volume | Date | Title | Pages of text | No. of illus-trations | Pages of bibliog-raphies | Pages of indexes |
|--------|------|-------|---------------|----------------------|------------------------|------------------|
| I | 1954 | *Introductory Orientations* | 248 | 36 | 50 | 20 |
| II | 1956 | *History of Scientific Thought* | 582 | 13 | 70 | 42 |
| III | 1959 | *Mathematics and the Sciences of the Heavens and the Earth* | 680 | 127 | 115 | 72 |

Bibliographies and indexes are presented in two scripts: in Roman, for non-Chinese sources, and in Chinese.

Mansel Davies adds:

The total physical compass of *SCC* to mid-1984 was: text pages 4,808; illustrations 1,202; bibliographies 1,285 pages: indexes 549 pages. This

15. Tao Té Ching (Dao De Jing), Ch. 31 (after Kao Heng [Gao Heng]), *SCC,* Vol. V, Part 6.

represents a minimum of two and a quarter million words of text, all written by Needham, but based on full-time support in particular sections from the collaborators. . . . It is not, of course, the size of the work which is relevant or necessarily impressive. It is the quality, the depth, and the enlightenment which characterize these volumes, and which form the greatest treasure in twentieth-century historiography.[16]

Needham confesses that when dealing with alchemy and early chemistry,[17] he and Lu Gwei-Djen 'almost despaired of ever finding our way successfully through the inchoate mass of ideas and the facts so hard to establish'. They got through finally by 'cutting great swathes of briars and bracken, as it were, through the muddled thinking and confused terminology of the traditional history of alchemy and early chemistry in the West'. They had to distinguish alchemy from proto-chemistry and 'to introduce words of art such as aurifiction, aurifaction and macrobiotics'.

Needham was also making a determined effort to publish books of 'less thickness, more convenient for reading'. By Volume III, he felt the books had developed a middle-aged spread. It was time to subdivide volumes if they could not be read comfortably in one's bath. Volume IV then was divided into three separate volumes, but that was not good enough. Volumes V and VI have so many subdivisions that Needham doubted whether he would see their completion within his lifetime. He was having to do this against a background in which it was highly difficult to predict the space he would require for the overall history. He found that 'our material is not easy to "shape"', and they were 'constrained to follow the Taoist natural irregularity and surprise of a romantic garden rather than to attempt any compression of our lush growths within the geometrical confines of a Cartesian parterre'.

The volumes that appeared were in themselves a source of special growth. They generated the East Asian History of Science Library, which in turn generated the East Asian History of Science Trusts, first in the United Kingdom, then in the United States and Hong Kong.[18]

16. Mansel Davies, 'Joseph Needham – The Erasmus of the Twentieth Century?', *New Humanist*, Winter 1986.

17. *SCC*, Vol. V, Part 2, p. xvii.

18. In the United Kingdom, the Needham Research Institute is the legal responsibility of the members of the East Asian History of

From 1942, Needham had been collecting materials for the library, a key source of revelation for the historian. He was aware that no Western library specialized in the history of science and technology in China. He and Lu Gwei-Djen brought together a unique collection of materials on the history of East Asian Science. Books seemed to hunt him out, as on the occasion when he went into a temple in the Wei Valley in Shensi Province, which was then the wartime centre of the University of Honan. To his astonishment, he stumbled across volumes of books hurriedly deposited there, including fourth-century texts on alchemy, gods and sages. Their existence was known only to a select few Chinese scholars. During the war years, Needham was able to purchase old wood-block Chinese texts, and have them flown to England. Although after Mao Zedong's victory in 1949 their export was prohibited, Needham was always favoured by the Ministry of Culture and Academia Sinica, so that the collection went on until about 1958.

Housed since 1991 in its permanent home at the Needham Research Institute, Sylvester Road, Cambridge, the library today has about 25,000 volumes and other materials, including a fine collection of photographs. It is the heart of a research centre with an international reputation.

When, in 1976, Needham retired as Master of Gonville and Caius, he formed the East Asian History of Science Trust, to which he donated his own library. The Trustees and the Cambridge University Press found him what obviously could be only temporary accommodation, first in a prefabricated building at No. 8 Shaftesbury Road, and then in one of Cambridge's older buildings at No. 16 Brooklands Avenue. The library could not function adequately there; life was difficult and cramped as space was far too limited.

However, at a dinner party in 1976, in a conversation with the Warden of Robinson College, Sir Jack Lewis, he suggested that the library might eventually have a purpose-built home in the grounds of Robinson College. Sir Jack was enthusiastic, and Needham wrote a brief for the building. He called in as archi-

---

Science Trust. Its present Chair is Professor Geoffrey Lloyd, Master of Darwin College, Cambridge. The sister-organization in Hong Kong is chaired by Dr Philip Mao, and another in New York has as Chair Dr John Diebold.

tect Christopher Gillet, his colleague as Junior (Buildings) Bursar at Caius. A detailed cost plan was prepared in 1981 by the quantity surveyors, Davis, Belfield and Everest. By 1984, the Trust was able to purchase the corner site between Sylvester Road and Herschel Road with funds for the central block and east wing supplied mainly by Tan Sri Tan Chin Tuan (Dan Sri Dan Jin Duan), formerly chairman of the Overseas Chinese Banking Corporation, who had met Needham in 1983 and was impressed by him.[19]

In October 1984, the Chinese Ambassador, Ch'en Chao-Yasam (Chen Zhoo Yaśam), laid the foundation stone. At the last moment, there was a danger that both the site and the funds would be withdrawn unless further finance was found. An anonymous donor came to the rescue. Later, it turned out to be Lu Gwei-Djen. That was not the only financial support given by her and Needham. Apart from reimbursement of out-of-pocket expenses incurred for the library, both gave their services free. Needham donated both his former Cambridge houses for the benefit of the Trust. From 1942 to 1955, he shared his salary as Sir William Dunn Reader in Biochemistry with Wang Ling, and from 1973 to 1979 he supported Francesca Bray from his own funds.

The south wing was built following generous donations from the American Kresge Foundation, on condition that the Needham Research Institute should find an equal sum. In June of that year, the K. P. Tan Foundation Ltd, Hong Kong, matched the sum required. The whole building, which houses the Institute that runs the library and looks after the *SCC* project, was completed in May 1991. The building was opened officially by the Cambridge University Vice-Chancellor, Professor David Williams.

The decorations of the Institute speak with a special voice. For example, fixed as a plate to each swing door is a square red seal *Wei Chung-Kuo K'o-hsueh Chi-shu Yung (Wei Zhonjguo Goxue Jishu Yung)* (For the use of the History of Science and Technology China). It appears also at the Head of the Institute's

19.  Other organizations providing funds were the EAHS Trust in Hong Kong, the Council of the Chinese Cultural Renaissance in Taiwan, and the National Commission for Science and Technology Policy of the People's Republic of China.

biannual *Newsletter*. It is an eighteenth-century seal made from a piece of amber marble, found in the summer of 1958 when Needham and Lu Gwei-Djen were on a return visit to Beijing. They were feeling good because the *SCC* project was running well and Volume III was in press. They agreed they had arrived at a moment for 'putting a seal on it'.

The seal they looked at was crowned with carvings of the four accepted pursuits of a classical scholar – the lute, chess, books and paintings. The weight and dimension felt right. The sealmaker decided to slice off the base on which there were other seal-marks. A seven-character inscription down a side of the scroll container reads *Pu k'o jih wu ju chun (Bu go ri wu ru jun)* (Not one day can go by without your company).[20]

20. 'Librarian's Notes', *Newsletter*, No. 3, January 1988.

# 9

# The Great Dream: *Regnum Dei*

*The Way to Heaven is Heaven Here and Now.*

1660 Quaker Declaration

Needham's *philosophia perennis* includes the conviction that evolution is a real phenomenon which Christianity has had to come to terms with. Evolution in all forms – cosmic inorganic, organic and social – is one long, continuous process. Therefore, the Kingdom of God on Earth is not some wild impossible dream, but something that has the whole authority of the evolutionary process behind it.

His feeling of revolt against a narrow Christian chauvinism enabled him to embrace Taoism (Daoism). As he often said, one of the most liberating influences in his life was to find in China that more than a quarter of the human race did not have the need to believe in a personal god or a creator. Although he stands resolute as a Christian believer, he objects to present-day cosmology linking the 'big bang' with the 'creation' in the Christian sense. He prefers a 'steady state' universe without a 'creation'. That is more in accord with Taoist and Buddhist theology. He had no idea of what the truth would turn out to be, but he hoped that it would not be provided by the big bang.[1]

He had self-doubts and fears. He often thought that one of the most awful things would be to die and then to find that the medieval-Christian conception of the universe was true. 'I think it would be terrible, absolutely terrible. I can't think of anything worse.'[2] And I recall the lonely moment when he said to me in a sad voice that in his conception of Christianity, 'I can see

1. *Sinorama Magazine*, January 1991.
2. Unpublished section of interview for *Sinorama Magazine*.

Dophi again.' There was always with Needham the hurtfulness of her years of suffering from Alzheimer's disease, so that she hardly came to recognize him as he looked after her and humoured her in his whimsical way. With her death, on 22 December 1987, a virtual part of his life was gone.[3]

His polarity was clearly expressed in reply to a letter from a friend, Frank McManus, who had got to know Needham while a student at Cambridge from 1945 to 1948. He had written to him on 16 September 1988 following Needham's BBC television interview with Ronald Eyre. McManus asked, 'Is "Ultimate Reality" an impersonal stream of destiny, as some Taoist traditions say, or is there a personal Supreme Entity, God the Father? I'm wondering if the former applies at the "earthly" level of consciousness, whilst God's grace raises His creatures thence to higher and spiritual levels when necessary?'

Needham, in his reply dated 28 October, stated that during his time in China he was chiefly attracted by the Taoist religion; hence his sobriquet, 'The Taoist of the Ten Constellations'. He knows 'there is all the difference in the world between the warm fatherhood of a personal god and the all-embracingness of an impersonal Tao'.

He went on: 'I am afraid that I do not know how to choose between these two forms of religion. I myself incline to the personal one, but it is for anthropological reasons; my forefathers were all Anglicans and that is why I am still one myself, but I can understand the appeal of the other.'

He stressed the

many extraordinary factors in our environment that have been pointed out for half a century past. For example, if the force of gravity was just a little bit stronger, or a little bit weaker, than it is we could not exist . . . the whole thing is summed up in the cosmological anthropic principle. . . . Thus it really does look as if there was some *preparatio evangelica* at work, and one might feel this speaks in favour of a personal deity as against an impersonal Tao. However, it is all rather a matter of 'one pays one's money and takes one's choice'.

As an overnight guest of Needham's at Caius in early 1976, I discussed with him the nature of religious experience and its

3.   Brian Harland, 'Joseph Needham', *Interdisciplinary Science Review,* Vol. 15, No. 4, 1990.

relationship to socialist experience. In this regard he was an idi-
osyncratic Marxist. His view was that religion was not necessarily the opium of the people; science might become that, meaning that entry into the world of the soul (that is, the experience
of the holy) was equally valid whether it were through religion,
science, art, or philosophy. He saw Christ as expressing the pro-
test of the oppressed. And religion, as the sense of the holy, was
as appropriate to man as the sense of beauty.

The Kingdom of God Needham understood as a realm of
justice and comradeship on earth to be brought about by man's
efforts throughout the centuries. But how soon could this be
made to happen? The more Christian a man wants to be the less
can he stay out of politics. 'Whether it's the wicked and geno-
cidal war in Vietnam, or the Orthodox Christians in the Soviet
Union, or the Bolivian tin miners in South America, everything
depends (it's a terrible judgement) on us, and according to our
actions, on us will be the judgement.'[4]

Thus Needham, concerned with rendering justice to a great
people, in 1952 was a member of the International Commission
for the Investigation of Charges of Bacteriological Warfare in
North China and Korea. Their report was affirmative. The
Americans were found guilty of employing biological warfare
during the Korean War. They cited, for example, the use of insects as carriers of plague and anthrax, and included the testimony of prisoners of war.

A press conference was organized on Needham's return to
London. It was held one morning in the Hotel Russell in
Bloomsbury. The hall was crowded: journalists from all over
the world were present. He was grilled for almost two hours.
There was barely a friendly question. I was impressed with his
performance. He did not evade any question. After the conference, he sought my advice. The *Daily X,* 'a reactionary paper'
Needham called it, wanted a special interview with him. What
should he do? He looked exhausted. I advised him to go home
and rest. He reflected, then agreed to do so, but hoped he would
not be described as a coward. Of course, for some time he was
regarded as a 'traitor' and ostracized by the British press and by
many academics. He was refused a visa to enter the United States.

4.  Joseph Needham, *Moulds of Understanding*, p. 265, London, Allen
    & Unwin, 1976.

And in the United Kingdom in certain authoritative circles he was once again treated as an 'Ishmael'.

Another aspect of Needham's search for the *Regnum Dei* was expressed by his alter ego, Henry Holorenshaw,[5] who described the seventeenth-century Levellers as the men who first saw the vision of the co-operative social commonwealth. In their pamphlet *The Light Shining in Buckinghamshire*, they laid down what 'honest people desire: (1) a just portion for each man to live, so that none need to beg or steal for want, but everyone may live comfortable; (2) a just rule for each man to go by, which rule is to be found in Scripture; (3) equal rights; (4) government judges elected by all the people; (5) a Commonwealth "after the pattern of the Bible".' The land to be the property of the whole people.[6]

The Leveller, Gerard Winstanley, was much favoured by Holorenshaw. Winstanley, for whom Jesus Christ was the Head Leveller, anticipated the criticisms of a religion that had made men think about a future life and so accept slavishly the evils of this. He recognized, also, the considerable part science was to play in the service of humanity. In a remarkable statement, he said:

If the earth were set free from kingly bondage, so that everyone might be sure of a free livelihood, and if this liberty were granted, then many secrets of God and his nature might be made public, which men do nowadays keep secret to get a living by; so that this kingly bondage is the cause of the spreading of ignorance in the earth. But when the Commonwealth's freedom is established, then will knowledge cover the earth as the waters cover the seas, and not till then. . . .

. . . to know the secrets of nature is to know the works of God, and to know the works of God within the creation is to know God himself, for God dwells in every visible work or body. And, indeed, if you would know spiritual things, it is to know how the Spirit or Power of Wisdom and Life, causing motion or growth, dwells within and governs both the several bodies of the stars and planets in the heavens above and the several bodies of the earth below, as grass, plants, fishes, beasts, birds, and mankind; for to know God beyond the creation, or to know what he will do to a man after the man is dead, if any otherwise than to

5.   In his book *The Levellers and the English Revolution*, London, Gollancz, 1939. (The New People's Library, XXI.)
6.   Joseph Needham, *Time, The Refreshing River*, p. 80, London, Allen & Unwin, 1943.

scatter him into his essences of fire, water, earth, and air, of which he is compounded, is a knowledge beyond the line or capacity of man to attain to while he lives in his compounded body.

The Levellers failed to introduce their desired economic system – a premature form of what Holorenshaw called 'socialism'. For Needham it has remained an inspiration. He has been a Christian socialist for almost seventy years, and he knows, to use John Polkinghorne's striking phrase, that he has 'not been given a blank cheque on a heavenly account'. But he is strongly aware that he was singled out to be given a blank cheque on a universal scientific culture account.

He is responsible 'for recreating a world of extraordinary density and "presence" ', as George Steiner comments. 'He is literally recreating, recomposing an ancient China forgot, in some degree by Chinese scholars themselves and all but ignored by the west. The alchemists and metal-workers, the surveyors and court astronomers, the mystics and military engineers of a lost world come to life, through an intensity of recapture, of empathic insight which is the attribute of a great historian but even more of a great artist.' He sets *SCC* beside *A la recherche du temps perdu*, for 'Proust and Needham have made of remembrance both an act of moral justice and of high art'.[7]

In the last year of his life, Needham's strongly held view was 'that by an extraordinary series of events modern science was born, and swept across the world like a forest fire. All nations are now using it, and in some measure contributing to its development. We can only pray that those who control its use will develop it for the good not only of mankind but of the whole planet.' This seems a re-statement of what he had found in the *Kuan Yin Tzu (Guan Yin Zi)*, an eighth-century Taoist book on the making of exceedingly sharp swords. The author wrote, 'Only those who have the Tao will be able to perform such actions – and better still, not perform them, though capable of performing them.' Needham commented: 'Mankind must learn this lesson, how to know, yet to refrain. The price is going to be survival itself, no less.'[8]

7. George Steiner, 'The Making and Progress of an Honorary Taoist', *Times Higher Education Supplement*, 1 June 1973, p. 15.
8. 'Greeting Card to Professor Abdul Rahman, January 1983', *Bulletin of Sciences*, December 1985/January 1986, p. 23.

He put this more extensively in a memorial volume to Louis Rapkine:

In the days when Bernal was writing, there was never any doubt that the natural sciences were essentially beneficial to mankind. The principal gravamen of the attack of these scientists on capitalism was that it prevented science from exerting its beneficial effects. Now the situation has utterly changed. There is widespread fear of science and no longer any conviction that its activities are always for the benefit of mankind, or even would be under socialism.[9]

As I come to end my story of Joseph Needham, I am most conscious that the 'mystery' that is Needham still remains. I recognize that my words can provide only an introduction to him. I am conscious of gaps I have left in my portrait, though I have given some impression of the tremendous physical and intellectual vigour that has gone into the expression of his creative originality.

There are what I can only describe as ideological lapses. When, on 1 October 1949, Chairman Mao Zedong, standing above the Gate of Heavenly Peace in Beijing, proclaimed the People's Republic of China, Needham wept with joy. Not only did he see the way clear for the building of a new society, but he was reacting, also, to Mao's statement that 'the Chinese people have stood up'. That linked China, Chairman Mao, the Chinese people, Needham himself and the English revolutionary tradition, for the song of the seventeenth-century Levellers was 'Diggers All Stand Up Now'.

Needham saw Mao as the person who made it impossible to introduce modern capitalism to China and to produce a country like Japan, and who insisted China could go straight to communism, that is, scientific socialism. He argued that 'the neo-Confucianism of the eleventh, twelfth, thirteenth centuries A.D. in China was a philosophy very congruent in its world outlook with dialectical materialism'.

In 1976, Needham wrote, 'in my belief, having been there a number of times since the revolution, the Chinese have made long strides to a society in which social classes have become entirely a thing of the past.' What then of the excesses of the Cul-

9. Joseph Needham, 'Science and Politics', in Vivian and Benjamin Karp (eds.), *Louis Rapkine*, pp. 91–3, Vermont, Orpheus Press, 1988.

tural Revolution? Needham believes that Mao went 'gaga' then. And he did not understand how the Chinese Party leaders could have made so barbarous an attack on the student demonstrators in Tiananmen Square.

It is embarrassing to read those words today, but they do make clear how Needham's love of China and his radical loyalty created his particular vision of reality.[10]

I am aware of the influence of his two wives. The gentle Dorothy, with whom he lived and worked for sixty-three years, to whom he owed a great debt of gratitude, as his companion and as probably, with Wang Ling, the only person who had read every word of the published *SCC* volumes, and whose improvements to the text have been too numerous to be cited. Her Chinese name was Li Ta-Fei, given to her when she worked in China from 1943 to 1945 as chemical adviser in the S-BSCO. The Chinese characters mean 'plum great, graceful'. She was an Honorary Fellow of Girton College, of Lucy Cavendish College, which she helped to found, and of Gonville and Caius, where she was the first woman Fellow and the Master's wife.

In 1989, Needham married Lu Gwei-Djen, whose name in Chinese characters was, 'of the Lu family, precious as the fragrant cassia tree'. When she first came to Cambridge in 1937, she studied for her Ph.D. under Dorothy Needham. She and the Needhams became close personal friends. She became for Joseph Needham 'the explainer, the antithesis, the manifestation, the assurance of a link no separation can break'.[11] Of him she said, 'Joseph has built a bridge between our civilizations. I am the arch which sustains the bridge.' At the lunch after the wedding ceremony in the college chapel, Needham commented, 'It may seem rather astonishing for two octogenarians to be standing here together, but my motto is, "Better late than never".'

Their married life together was short. She died within two years of broncho-pneumonia. Her death affected him much more than that of Dorothy. Professor Gwei-Djen Lu-Needham was Associate Director of the Needham Research Institute, a founding Fellow of Lucy Cavendish College, and Emeritus Fellow of Robinson College.

10. 'On the Death of Mao', *New Scientist*, 16 September 1976, p. 584.
11. Attribution in Joseph Needham, *The Grand Titration*, London, Allen & Unwin, 1969.

Needham had set up The Lu Gwei-Djen Memorial Charitable Trust, which has three major objectives: (a) completion of the *SCC* Project; (b) establishment of studentships to support young scholars from China with interest in the history of science, scientific thought, technology or medicine in China; and (c) strengthening of the Endowment Fund of the Needham Research Institute. The trustees are Professor W. Y. Liang, Dr Hsing-Tsung Huang, Professor Geoffrey E. R. Lloyd, Mr W. Brian Harland, and Mr Kenneth G. Robinson.

Now the ashes of both Needham's wives lie buried beneath the same wuthang tree (also known as the phoenix tree) in the approach to the Needham Research Institute. Needham had told Professor Ho Peng-Yok, his successor as Director of the Needham Research Institute, that he never expected to live to the age of over 90. When Professor Ho suggested that it was due to his genes, he denied it, saying it could not be true because his mother died in her seventies and his father in his sixties. He attributed his longevity to ginseng. He started taking one pill a day in the 1960s, but in recent years he cut it down to one pill a week on the advice of Lu Gwei-Djen.[12]

Needham, made up of many elements, was a supreme bridge-builder. He was not only a witness to a great truth, he was also a creator of that truth. He was a twentieth-century colossus, and his Great Dream must inevitably come to be realized if humankind is to survive. As a close colleague of his said, 'His Maker must be very pleased with him!'

12. *NRI Newsletter,* No. 13, January 1993.

# Appendices

The reminiscences by Dr Wang Ling are made up of two contributions I have linked together and edited slightly. They consist of a long personal statement dated 4 April 1994 in reply to a request from me, and his opening address to the fifth International Conference on the History of Chinese Science and Technology, held in San Diego, USA, in 1988.

Dr Wang Ling died suddenly on 6 June 1994, age 76, in Nantong, Jiangsu province, where he was born on 23 December 1917. He had a great admiration and deep affection for Joseph Needham, whom he regarded as his true mentor.

Maurice Goldsmith

# Reminiscences of Joseph Needham

*by Wang Ling*

## *Joseph working*

When I first arrived in Cambridge from China, I went to Room K1 in Caius Court at College. I was astonished to see Joseph himself carrying hundreds of boxes full of reprints across the court to his car. These reprints on chemical embryology he had tied up in bundles in their original order to send to the Biochemical Institute of Academia Sinica in Shanghai.

One may think the old reprints would have only historical value, but some of them still had living interest. For example, some years later Joseph received a telephone call from Cambridge railway station. It was from a stranger, who had just arrived from France, asking for a reprint of a reference named in his *Chemical Embryology*. Joseph said, 'I cannot see you now, and I do not have that reference any more. I sent it away some years ago.' But the caller would not take no for an answer. He said, 'Please, I would like to invite you to a meal just the same. I have specially travelled from France.' At the meal he produced a handsome cheque saying, 'This is a token of my gratitude. As the managing director of an egg-producing business, I have made millions by producing many times more and better quality eggs by following your note.' So Joseph graciously accepted his cheque, putting it in his Trust for funding his *SCC* project.

Joseph, as a rule, when working would not let anyone disturb him, no matter how famous the visitor. One time the distinguished statistician and Darwin Prize-winner, Sir Ronald Fisher, knocked and immediately opened the door, one foot already in the room. Joseph said, 'I'm frightfully busy.' It turned out that only a few hours earlier, he had asked Sir Ronald's favour to let him use his room in College to put up his Chinese scientist visitors. Sir Ronald, also a most aggressive type, in order to drown the noise of Joseph's typewriter, shouted back, 'I just came to tell you that I can put up your visitors. You asked me and I say "yes" ', and off he went abruptly, without waiting.

Another example was when his old friend Julian Huxley, the first Director-General of UNESCO, having come all the way from London, phoned from the porter's lodge. Joseph again said, 'I am frightfully busy. You come without an appointment. I am afraid I can't see you.' So Huxley went back to London without seeing him. In spite of these incidents, these three remained very good friends.

Anyway, this is how he gets jobs done – once he starts he never stops. He never makes any difference between important visitors and unimportant ones. When he starts working he does so non-stop. He does not need the help of either a typist or a secretary. He composes his manuscripts directly on his typewriter, using only two fingers and typing at terrific speed. The typescript was his first and final draft to be sent straight to the Cambridge University Press to be printed. The only time he would pause, and then very briefly, was to say 'Thanks awfully', when the porters brought in his letters, and immediately he would go on typing.

He has some particular habits in working non-stop. The Chinese have a proverb to describe a hard-working scholar reading books all the time, even reading while travelling on horseback. Needham travels by train, always buying a first-class ticket, not because of any snob value, but because only the first-class has empty compartments where he can spread his books and manuscripts around, jotting down notes, in preparation for the next section of *SCC*. Even when travelling by car, while driving he always discussed some topics of his book. However, there was one exceptional occasion when he did not discuss the book. He was driving at top speed on our return journey from a visit to Oxford, for he was engrossed in thoughts of his wife, Dorothy, who was in hospital in Cambridge about to undergo an operation. Suddenly, he noticed the passenger seat beside him was empty. As one would expect, a Chinese was too polite to ask him to stop the car in order to secure the latch on the door which was not properly closed, so I had fallen out from the fast-moving car. Joseph was upset beyond description, but I have survived to tell the tale.

Another time he took me to Oxford where he lectured to a Summer School during the holidays. We considered this a holiday break, but while he lectured from his manuscript he was proof-reading at the same time. On another occasion he took me to visit the Greenwich Maritime Museum. In preparation for Volume IV, Part 3 of *SCC*, on the History of Nautical Technology, he made use of this holiday break to discuss with their naval historians the subject of ship-building, in particular comparing the early appearance of the stern-post rudder in Chinese history with that of the West. He went to Greenwich expecting to pick the brains of the ship-building experts, but it ended up with those distinguished historians learning a tremendous lot from him. With their eyes popping out of their heads, they cried in admiration, 'How wonderful the ancient Chinese!'

None of Joseph's time is wasted. Even if he had only a few minutes to spare he would use the time to cut cards. This he himself called 'knitting'. These cards were originally menu cards from the tea house across the road, where he sometimes took me to tea. On these cards he would jot down some thought or idea that came to him while having tea, or perhaps some interesting remark he may have overheard. These menu cards he cut and filed straight away in their proper places, as he always said, 'A place for everything and everything in its place.' He filed his works most methodically, be they pages of translations or teahouse cards.

The tea-breaks were for him also working sessions. At such times he often had the company of great scientists in various disciplines, and he would often make unexpected new discoveries over a cup of tea. For instance, I still remember Professor Gray inquiring about the rain-gauge history in China and Joseph jumped up to search for a reference to the first rain gauge accidentally preserved, believe it or not, in a mathematical book of the Mongol period. While the volume of the rain-gauge vessel was being calculated, the teacups remained full, the tea now cold and undrinkable. Thus Professor Gray left K1 staircase in Caius College without tea, but with the knowledge of our unexpected discovery that the Chinese had produced the first rain gauge in the world.

A tea-break once brought Needham and me to the outskirts

of Cambridge. I remember we came to a deserted RAF war-time airfield. There, in an old wooden hut, I listened attentively to his conversation with Dr Martin Ryle, who later became a Nobel Prize-winner for his pioneering work in radio astronomy. Joseph recalled Chinese continuous historical records of novae and supernovae, including one appearing on the Chinese oracle bones in the second millennium B.C. and the Crab Nebula in A.D. 1057, recorded by a Chinese astronomer. Ryle showed us his structure of wires which he designed as the first radio telescope of such type, set up in 1946. It enabled him to make investigations of the radio emissions from the most distant radio stars, some of which are too faint to be photographed by the 200-inch optical telescope at Mount Palomar in California. Thus, Joseph's oracle bones novae of the second millennium B.C. can no longer be seen by the world's largest telescope with all its sophistication, but can find living witness in Martin Ryle's makeshift shack with its long-stretching wires in a rugged, disused airfield near Cambridge. So the tea-break brought Joseph and myself as early as 1947 to see that derelict aerodrome, where Ryle opened up a new branch of learning, and Joseph linked the Chinese historical record with Ryle's modern observation. This was the hard-working, highly perceptive Joseph with his cups of tea.

Occasionally, only a very few times, he stopped working altogether on *SCC* for a few weeks. For instance, once he stopped to write an annual review of a distinguished scientific journal. All the world-renowned scientists chose him to write this. He had to read all the important papers published in that journal for that year. Eventually, he sifted them out and narrowed them down to the 8,000 best papers. Then he had to digest these 8,000 into one review. He was the right choice. He can read fast and select important points and weave them all into one review.

Joseph was engaged full-time in the history of Chinese science and technology and medicine, but at the same time he was still holding his readership in biochemistry and teaching three special courses. As for his teaching, every year he gave more new material and was not merely repeating old lectures. In doing this, he always taught very dutifully. He had a special habit of having his first course in the morning, so the previous evening he went to the lecture room to write his notes on the blackboards. Cambridge blackboards have many layers on the walls working on a pulley system, so that one layer can be pulled down over another.

He wrote rapidly on the blackboards, so that he would not have to waste time during the class. However, on one occasion he had a disaster. He always made sure that the cleaning woman did not touch his blackboards, but one day a new woman unwittingly wiped out his lecture notes. He was furious. However, it was not as disastrous as it might have been because he remembered everything and gave the lecture without notes, off the top of his head. The only problem was that it was more difficult for the students. He was only able to do this because of his fantastic memory, and also because when lecturing he spoke extremely fast.

At his breakfast table every day, Joseph had the latest scientific journals in his own field, and if he found anything new and relevant he would jot down notes to add to his lectures. Breakfast time was his working time, too. He was reading everything and underlining and making notes in the margin, often discussing articles with his wife Dophi. He followed up all the newspapers and journals with astonishing absorption. On one occasion, when he was Master of Caius, Joseph invited me to High Table. Next to me was Dr Crick, the discoverer of DNA and Nobel Prize-winner, who had invited a very bright young American scientist to the meal. The latter was raising questions on new developments, trying to show off to Dr Crick and the Master. Whatever new research was mentioned Joseph had not only already read about it, but also himself asked the young man questions on the subject. I was astonished at how quickly Joseph answered the questions put to him, and again put new ideas into the young man's head.

One day I was invited to have breakfast with Joseph and Dophi. There was another guest, a Russian scientist. In the middle of breakfast, the Russian asked a question. His small science book had been translated into English by someone he believed was using a pen-name. He had for many years been trying to find out the actual name of the translator and it still remained a mystery. Joseph just stretched out his hand to the bookcase behind the breakfast table and fished out a book. To our surprise, he produced the very book! He himself had been the translator many years before in his undergraduate days.

He knows so many languages. He is practically bilingual in

French. His German is extremely good. In 1943, he visited our institute in Lichuan (Lizhuan). Down the hill from the institute was the T'uang Chi (Duang Ji) University, which specialized in medicine and had many German professors. He gave lectures there in German off the top of his head, and he answered in German many questions asked by the German professors. He also knows Italian well. In 1956, when he was surrounded by Italian journalists on the occasion of the International Congress in Florence and Milan, he answered their questions in Italian off the cuff. He can also read and speak Greek, and very often went to Greece for a holiday. He can not only write but also speak Chinese. When in China in 1943, he deliberately employed a driver whose only requirement was to be able to speak standard Mandarin Chinese, with no knowledge of English. This was the essential qualification of his driver, making no mention at all of driving skills. In this way, he could learn to speak Chinese with his driver while visiting various institutions and universities. He insisted on writing all his reference cards in Chinese, writing every character himself, even if the stroke order was occasionally incorrect.

It was also in 1943 that he came to give a lecture at our Institute of History and Philology at Lichuan. Dr Fu Ssu (Si) Nien was in the chair. Our lecture room during the war doubled as our dining room. With tables moved into one corner it became a lecture room. It was summer and Joseph came into the lecture room wearing shorts, carrying all his books. He found only a tiny table in front, with not enough room for them all. Without a word, he marched to the corner and lifted a heavy, square cedarwood table upside-down and carried it high above his head to the front of the lecture room. He put all his Chinese books on this table. As he lectured he fished out all the unpunctuated sentences from these old books and immediately read out fluently, in classical Chinese, his selected quotations. We were all taken aback by his astonishing ability in reading classical Chinese.

When I recalled this incident many years later during my visit to Cambridge in 1989, Joseph said from his wheelchair, 'I was very spry in those days.' He was referring to his action in lifting the heavy table so easily, and was not blowing his own trumpet about his impressive reading of classical Chinese, although everybody else was impressed by his ability.

To add to these examples of his ability in writing Chinese, I must mention that once I received a letter from my mother asking me to return home. Joseph took the trouble to answer her in a letter written in Chinese. I made some corrections for him and he immediately copied the whole thing and sent it away. I am now trying to find out from my brother and other relatives whether the letter is still in existence. I still remember when he came to the point in the final paragraph, he copied it very nicely with a contented smile: 'Raw rice which has already been cooked cannot return to its original state', meaning 'The writing of this book is almost accomplished. You will one day be proud of your son.'

## Helping people

For some time prior to 1956, Joseph shared his own salary with me without telling me. Only much later I found out that he had done this. Then Professor Tzu K'u Chan (Zi Gu Zhan), who was head of a Chinese delegation taking part in an international congress, heard of this. Knowing Joseph wanted to keep me with him, on returning home he asked the Chinese Government to give me a grant. This they did and I was holding this grant until I went to Australia in 1958.

In 1956, the International Congress on the History of Science held in Florence and Milan had very strong delegations from the United Kingdom, the United States and the USSR as well as China. But only Professor Tzu K'u Chan (Zi Gu Zhan) was elected First President, in addition to the leader of the host nation, Italy. I think Joseph must have played a certain part.

In 1948, Professor Tzu K'u Chan was elected a member of the Commission for the History of the Social Relations of Science at the International Union for the History of Science. This small élite group consisted of distinguished professors, such as Gordon Childe, author of *The Dawn of Civilization*, Benjamin Farrington, author of *The History of Greek Science*, and Professor Eberhard, author of *The Rulers of China*, an outstanding professor of Chinese history at Berkeley.

In helping young Chinese scholars, the most notable was Huang Hsing-Tsung (Xing Zong). He got a scholarship to Oxford, presumably with Joseph's help. Then Ts'ao T'ien Ch'in (Ze Dian Cin) took Huang's place. Afterwards, he won a British Council Scholarship to work at Cambridge, where Joseph offered him his guest room in his own home in Owlstone Road. Ts'ao chose to do an undergraduate course, which was very difficult as he had to compete with the best British students. Normally, only a very few gained first-class honours, and in spite of all the competition Ts'ao succeeded in doing this. He went on to do a brilliant Ph.D. As a result of his achievements he was elected a bye-Fellow of Caius College, which meant he would have automatically become a Fellow had he stayed on. This was a great honour as he was the first Chinese ever to be elected to a Fellowship of a Cambridge college. Joseph jumped three feet in the air in his boundless joy when he heard this record-breaking news, and immediately wrote to the President of Academia Sinica in China, Kuo Mu Ru (Guo Mu Ru). Ts'ao later became director of the Institute of Biochemistry in Shanghai and was world-famous in his field. He was one of the major researchers working on synthetic insulin and enabled China to forge ahead of many other countries in this discovery.

Another scientist, a female assistant in Joseph's Sino-British Scientific Co-operation office in China, was given a British Council scholarship to work with Dophi in biochemistry. After she returned to China to continue her work, Dophi helped her to buy equipment costing many hundreds of pounds, which she paid for from her own pocket – naturally, with the encouragement of Joseph.

During Joseph's years in China, he helped scholars in institutes and universities in many ways. For example, he would read many papers he thought were worth publishing and would recommend them to distinguished learned journals. These papers by younger, as yet unknown, scientists would not otherwise have come to the attention of the editors of these journals. In this way, he must have helped hundreds of scientists, not only in biochemistry, but in many other disciplines. For example, I know one person, Wei Te-Hsin (Dexin), a statistician, who had several papers published through Joseph's recommendation and who later won a British Council scholarship.

Joseph persuaded the British Government to establish British Council scholarships with stipends equal to the earlier Boxer Indemnity scholarships. This caused the Chinese Government to conduct a nationwide public examination to choose the top people in various disciplines to benefit from these scholarships. Later, nearly all these people returned home to become leading experts in their own fields. All turned down offers of permanent university posts in England to serve China.

When Joseph was head of the Sino-British Scientific Co-operation office in China, he made very many friends of Chinese people whom he met at universities and institutes wherever he went. With his fantastic memory he never forgot the names of those he met, however briefly, and wrote them down in Chinese. Whenever he met these people again he could always put the right name to the right face. He visited libraries and always noticed when certain important books were lacking. He wrote to the British Government asking them to buy many of these books, for instance, the large ten-volume *Oxford English Dictionary* for the Fu Ssu (Si) Nien Library, now in Taiwan. Thus, he continued helping Chinese friends in need. China was so poor and isolated during the war and needed help. Since they could not show their gratitude by sending expensive gifts, and they also knew that he would never accept expensive gifts, many Chinese friends gave Joseph small gifts, such as silk ties. Even now he is the proud possessor of more than 300 ties! When he moved house from Grange Road to his present home in Sylvester Road in 1989, we helped him pack up his tie collection, among other things. Institutes, such as the South West University, sent him scrolls inscribed with the names of all the professors. Recently, Yang Chen (Zhen) Ning, the Chinese Nobel Prize-winner, told me that when he visited the Needham Research Institute in Cambridge he saw his father's name, Yang Wu Chih, who was professor of mathematics, on one of these scrolls.

## The Dickinson Lecture

Joseph has a special gift of learning a new subject, not his own, by rapidly reading some of the experts' work on this particular subject. For instance, in iron and steel technology, he became an expert himself in no time. Sometimes, he would invite an expert in the new field to tea and discussion, sometimes he would just read. He has the gift of being able to condense a big topic, of which he has so much knowledge, into one lecture without leaving anything out, as he did in the Dickinson Lecture. He made many discoveries which others had missed.

I still remember when I accompanied him on the occasions when he was invited to give important lectures on engineering, attended by many famous engineers. We were given red-carpet treatment and invited to fine meals. I learned that these engineering professionals were astonished that he knew so much.

Later, Heffer's published his Dickinson lecture in book form, and this shows a classic example of the way he digested new material and included all the important facts in one small book. He was the first to make so many discoveries about Chinese achievements in ancient times. Then comparing East and West, he demonstrated that the Chinese were the first.

## Religion

Joseph was a very religious man. Almost every Christmas, he went to the special Christmas Eve midnight mass at Thaxted Church. The priest at this church was noted for his progressive thinking. This church is usually crowded, with a full choir and even an orchestra, and is very well known for its music. On one occasion, Joseph was specially invited to give the oration at the memorial service for a very important person. We heard him dictating the text of this sermon on his dictaphone. On the appointed day a chauffeur-driven car arrived to take him to Thaxted.

In September 1989, when he decided to marry Gwei-Djen, he planned to keep it very quiet and have a very small wedding with just a few close friends attending. However, the information was leaked and word spread literally around the world. Bouquets of flowers arrived by Inter-Flora from every continent, including China and Japan. There was even a special floral tribute from the Archbishop of Canterbury, Dr Ramsey. They were married in the Chapel of Caius College by the Dean of the College, and afterwards had colleagues and friends to tea in the Upper Combination Room, presided over by Professor Macpherson, the Caius College president, and Lord Lewis, the Provost of Robinson College.

## Gwei-Djen

In 1989, Joseph still retained his old room K2 in Caius Court. From time to time, he dined at High Table to keep in contact with his colleagues. As he was confined to his wheelchair, we used to assist him from K2 to the dining hall.

In K2 we saw a very beautiful photograph of Gwei-Djen, taken by a professional photographer. She was looking very glamorous. It was on his mantelpiece side by side with another photo taken of her when she arrived at Cambridge in 1937. Joseph said to us, 'You wouldn't think it was the same person, would you? When she first arrived butter wouldn't melt in her mouth.'

He often commented that Gwei-Djen knew many idiomatic phrases. I asked him, 'Give me an example.' Joseph said, 'She often uses the expression "making heavy weather" of something.' Again, she frequently said she was "living on borrowed time", meaning she had several times been seriously ill and had had one lung removed.

Another time, when Joseph was Master, we came from an upstairs room and were astonished to find a black shadow coming out of the dark entrance hall of the Master's Lodge. It was Gwei-Djen, with her hand holding her right side. When Joseph and Dophi came back from the cinema, they took her immediately to hospital. This quick action saved her, as she had a perforated appendix.

Joseph, Dophi and Gwei-Djen were all very fond of hot crumpets. It was their favourite tea-time dish. Joseph would often toast his crumpets by his electric fire in his study so that no time would be wasted and he could go on working and writing. Sometimes, Gwei-Djen teased Joseph, saying, 'Can you cook?' Joseph would reply, 'Yes, I can.' She then would ask, 'Tell me what you can cook', and Joseph would reply, 'I can boil an egg!' The reason for this was that he wouldn't have to waste time watching it, but just leave it in boiling water while he kept on working.

Dophi and Joseph were for many years, in fact since they were first married, looked after by 'Auntie Violet', who always gave them a very good breakfast. Even after he became Master, he kept Auntie Violet on, although she was then over 90, and employed her to do only a few light chores, just to make her feel wanted. Of course, the College supplied a maid for the regular work at the Master's Lodge.

## Joseph and our boys

When I was visiting Cambridge in 1970, our boys were about 6 and 7 years old. Joseph invited us to the Master's Lodge several times. Sometimes we had a meal there, sometimes tea. In the long dining room, the floor was highly polished and the two boys took the opportunity of sliding on the slippery surface, as if roller-skating. At the end of the room, in danger of being crashed into, were cupboards with glass doors. I tried to stop them, and Joseph said: 'Ah, you can't expect active boys to be always cooped up inside.' So we encouraged them to run around the garden.

In the garden stood a ch'ang. Ch'angs are small stone posts carved with Buddha's name or with Buddhist incantations. They are usually scattered by the roads in the Chinese countryside. This particular ch'ang, with a number of others, had been planted in North Wales to make the landscape resemble the Chinese countryside. This was to create the background scenery for the film *The Inn of the Sixth Happiness,* which was shot in North Wales, starring Ingrid Bergman. After the filming, Joseph had travelled to North Wales and bought this ch'ang and transported it to grace his garden.

While the two boys were playing, they climbed onto the ch'ang, and it collapsed with a resounding crash. We saw Max lying under the ch'ang. While we were scolding the boys and apologizing for destroying his precious garden decoration, Joseph was rushing to see if Max was hurt. Suddenly, Max jumped up without a scratch but the ch'ang was in fragments! Again we made a profuse apology for the boys. Joseph smiled and said, 'We should rejoice that the boys are not hurt. The ch'ang is only a plaster-of-Paris imitation. It will be O.K.' And, indeed, a few days later, all the pieces were fitted together again and the ch'ang was restored.

Joseph's influence on me was even extended to our boys, in whom he took an interest when they were very young. For instance, he and Dophi took them to see the old steam locomotives at the Nene Valley Railway Museum near Peterborough. Joseph had the history of steam locomotives since the Industrial Revolution in his head, which he imparted to them. Many years later, when we were showing visitors the sights of Canberra, including the oldest locomotive in Australia exhibited as a museum piece at Canberra railway station, the visitors were astonished at the detailed history the boys were able to add to the conversation. Little did they know that it was information remembered from their childhood visit with Joseph.

Our younger son, Stephen, now aged 30, has been working for some years at IBM on software projects. At home, he continues to use his software skills inventing a number of computer games which impressed the American firm, Electronic Arts. Now he is planning to go to Los Angeles to a computer-game developers' conference. One would never have imagined that his interest could be traced back many years to the time when he was very small and had been given a number of science-fiction books. At Cambridge, at the age of 6, Joseph had given him and his brother a number of books by Ursula LeGuin. He took the trouble not only to inscribe the flyleaf of each volume with a short note about the author, but also added the titles of companion volumes, in their correct order, which he could not get immediately, so that the boys could try to get hold of them themselves later on. This they eventually did. It was on the contents of these books that Stephen based his computer games, such as his 'Demon Stalker'. A few years ago, Stephen and his co-worker came to Cambridge and made a point of visiting Joseph when he was still living in Grange Road. Joseph took the trouble to show them his newly built Institute and how his 'engine room' worked, with all his cards and files and cross-references.

## *My birthday poem*

My birthday poem for Joseph was originally written in 1989 while I was at Cambridge. I wrote it in Chinese calligraphy, framed it and presented it to Joseph. What I enclose here has been slightly altered to bring it up to date concerning the year and his age. It was translated by my wife Ruth and me while in Nantong in 1993.

## Poem celebrating the 92nd birthday of Joseph Needham

*Congratulations on your 92nd birthday. Adding one more counting rod we celebrate the rare Needham.*

*These years' remarkable achievements are made all the more astonishing with the record of two beautiful beings[1] and from*

1.  The two beautiful beings: The first beauty is Dorothy Moyle. Dorothy and Joseph were the first and only couple to be Fellows of the Royal Society, breaking the record in the history of the British Royal Society. The second beauty is Lu Gwei-Djen who gained much distinction and outstanding honour. She was successively

*unattainable laurels.² With admiration and gratitude my love and affection for you have been deepening with the years. Should this be compared with the Cam River, it would be like its tumbling white waves moving on and on, never to end.*

In the dining hall of Caius, as in all college halls, there proudly hang the portraits of eminent people who over the years had

---

Foundation Fellow of Lucy Cavendish College and then of Robinson College. To be a Foundation Fellow is indeed not easy, but less so to be a woman Foundation Fellow. For an English person to be elected like this is already difficult, for a Chinese it is even more difficult. This indicates that she was recognized for her collaborative work with Joseph.

2. The four unattainable laurels are as follows:

   1. Joseph attained both the FRS and FBA in one person. Throughout the world there is no equal. This is the first unattainable laurel.

   2. In 1943, Joseph became the director of the Bureau of Scientific and Cultural Co-operation between China and the United Kingdom for about four years. He bought books and scientific equipment for China, established the British Council Scholarships to develop and cultivate the talent and encourage the cultural exchange between the two countries and never spared any effort. Now after fifty years, one can meet people who sing his praise everywhere. This is the second unattainable laurel.

   3. From 1966 to 1976, Joseph was the Master of Caius College. Since then the professors and porters have been treated as equal colleagues by the Master. Even now, sixteen years later, everyone still remembers his democratic attitude and speaks highly of it. This is the third unattainable laurel.

   4. The fourth unattainable laurel: Joseph established the Needham Research Institute not only for the completion of his unfinished work, but also for the permanent preservation of valuable materials he collected for generations to come. He wrote the monumental work *Science and Civilisation in China* in several tens of volumes, to promote mutual understanding among the different cultures. So remarkable is this work, so lucid in style, so beautifully written, that many distinguished persons, scientists and statesmen alike, proposed Needham to be the Nobel laureate for peace and literature.

graduated from the College. For his portrait, Joseph chose to be painted wearing a blue Chinese gown as a symbol of 'recognition of the high level of Chinese achievement'.

Many Chinese scientists used to offer to help Joseph in many ways. Occasionally, his regular helper, Dr Yuan Jun, was absent and he would phone my wife and me, since we had known him for many years and he had got used to us. We helped him with tasks, such as putting on his shirt and good suit, so that we could wheel him in his wheelchair to the College to dine at High Table. When we were leaving Cambridge he said, 'I shall miss you dreadfully.'

# Needham's closing days

*by Stanley Besh*

I'll take you from Wednesday, 22 March 1995, two days before he died. The usual care was given in the morning when I got him up, I noticed at the breakfast table that he wasn't eating too well; in fact, he had great difficulty in eating, but with help I was able to assist him, but he wasn't eating easily – he was a man who loved his food. Because Joseph had difficulty in swallowing, drinking was also a problem.

However, he showed some interest in what post had arrived and wondered if there was anything from abroad. This morning there was not. There were one or two pieces of junk mail. He watched a bit of early morning news, but then I switched off the TV and he listened to his classical music, which he often had on during his breakfast.

After breakfast, I started to help him get ready for going over to the Institute at 12 noon, part of his routine every day. This process took some time every day because of hygiene and toiletries that needed to be attended to. Joseph always helped himself as much as he could in his own way. He always eagerly looked forward to going over to the Institute, and when he returned at 5 p.m., I would ask him if he had had a good day, to which he would reply, 'Yes, I had a good day at the Institute.'

He would look at the evening newspaper immediately upon arrival. I then took him into the bedroom, as I always did, and undressed him and put on his *chang gua* (Chinese gown), in which he was much more comfortable, a routine which he had carried out for many years.

I decided that we should have an early dinner, in fact it was supper/high tea, and I prepared him steamed fish with a poached egg, though he was not able to take very much of that down, because of the difficulty of eating. He liked ice-cream very much, always had done, and would eat it morning, noon and night, if possible. Tonight was no different, and he had ice-cream with a chocolate sauce that I made.

After this I asked him if he would like to go down to the study/lounge, where he always liked to be, usually to watch Channel 7 on Cable TV because he was able to watch all the news, particularly the news from China. But then to my surprise (when I was clearing away the dishes, I would pop in to see how he was), he had manoeuvred himself in his wheelchair over to his large desk, where he normally sat to attend to domestic paperwork. Again, much to my surprise, he was tidying the desk, putting papers in order it seemed, and elastic bands round cards, etc., placing his overseas stamps in envelopes, and was quite happily doing this for about an hour.

Of course, while he was doing this, I intermittently went down to his side to give him water, or a drink of some sort. Later I went down again, to find that he had manoeuvred himself back to the viewing position, to watch the television again, and I sat with him for a time. I looked at him occasionally and thought that he didn't look too well (he often had not looked well for the last week or so). So I suggested that he had an early night, and he agreed; at 10 p.m. I wheeled him back up to his bedroom, prepared him for bed, made his drinks and helped him get into bed and made him very comfortable, with the usual farewells of many nights in the past, 'Goodnight, Joseph, God bless', and he would reply 'And God bless you, my dear boy'. That was rather nice. As usual, I would go into him throughout the night; this Wednesday night, though, was a little different and I went in a little more often because of the change in his state of health during the evening.

On Thursday morning, he did seem somewhat improved after his night's rest, which had obviously perked him up a bit. I used to joke with him, 'Your batteries are charged up for another day, Joseph', and he would laugh at that. He decided that he was going to have his shower after breakfast this morning, so he had a quick wash and then I moved him to the breakfast table for his prepared grapefruit (a regular choice), orange juice, his vitamin tablet and medication, together with his cereal, which with my help he was able to take quite a lot down. This was followed by his toasted roll, with his favourite thin-cut marmalade – I often caught Joseph digging his long-handled spoon into the jar for spoonfuls of marmalade, which of course I turned a blind eye to, particularly as he was enjoying himself so much.

After breakfast, I prepared him for his shower and hair-wash, his thick white hair was quite long now, and I teased him, saying that he looked like a mad professor who was going to blow the world up – Joseph would go into hysterics when I said things like this. He was quite delightful that morning and in good form. When, after his shower, the nurse had attended to the dressing on his foot, which had to be done two or three times a week, I helped him to get ready to go to the Institute.

I said to him, that it was still early, about 11.30 a.m.; perhaps he would like me to push him round the garden, because the spring flowers were everywhere. It was clear that he was happy to do this. I wheeled him out to the gate entrance on the driveway, where he was able to see closely the daffodils, grape hyacinths, scillas and blue anemones. I said that the latter reminded me of his blue eyes, and he smiled. He then wanted me to push him as near to the plaques where the ashes of his two wives rested, and I quite calmly said to him, 'Is that where you're going to go, Joseph?' 'Oh yes, oh yes,' he replied. We watched the squirrels and the various birds that visited our garden. Joseph had a bird-table outside the flat window where he could watch when he himself was eating. The squirrels would come and help themselves to the birds' food; in fact, last year we had two families of squirrels, twelve in all, sharing the bird-table. We continued our walk down to the Institute to be there for 12 noon, where Tracey (Mrs Humphries, née Sinclair) had arrived.

When he returned at 5 p.m. on Thursday evening, again I thought that he did not look too well, so I took him off to the bedroom and undressed him. Then I put him in his *chang gua*, which he liked to wear in the evenings. I prepared a light supper for him, and as I went to put it in front of him, he slumped over a little bit and his arm dropped down beside his wheelchair, I thought it might be tiredness, but thinking further I thought he may have had a mini-stroke, so I gently put his arm on to a pillow which was in the chair, and sat him up a bit, called Dr Cartwright, who came straight away. By the time the doctor arrived, I had put him into bed. The doctor agreed that he may have had a mini-stroke, but it was hard to tell; there might well be another one in the middle of the night, so I was to keep a vigilant/strict eye on him. Once or twice during the evening I went into him, and he seemed to be a bit more cheerful, and I was able to talk with him.

On the Friday morning, after a fairly good night, he did not seem too bad, but he was certainly not his usual self, so I decided not to shower him but give him a blanket-bath instead. I prepared the usual breakfast, and he sat at the table, with his classical music playing in the background. I was thinking it would be rather sensible if I could keep him at home that day, although the Institute was his life and he was never prevented from going, and he would invariably insist on it. This put me in a little dilemma, so I said to Joseph that I would be back in a moment, and that I was just going to pop over to the Institute.

I normally spoke to Angela King about situations such as these, but on this occasion it was John Moffet, because Angela was away on annual leave. I was anxious that the Institute staff would not think that I was bringing Joseph over in a very poor state of health, which they knew was not the case, and they perfectly understood that if Joseph insisted on coming down, there was really nothing I could do, except abide by his wishes, though if I could talk him into staying at home, I would do so.

Much to my surprise when I returned to his flat – Joseph of course was still at the breakfast table, he was managing to drink his coffee. I said, 'Joseph, wouldn't it be rather nice if you had today off?' He looked at me quite wide-eyed. I continued, 'Because it is Friday, why not make it a long weekend, and charge our batteries for Monday.' 'All right,' he said, 'I'll stay at home.' This pleased me immensely because I knew then that I did not have to rush him at all this day. I then called the surgery; the district nurse was due to visit that very morning and I helped prepare for her impending visit. In fact, the nurse arrived just before noon, as did the doctor. Joseph's co-carer, Duncan Manson, arrived at this time.

The doctor suggested it might be a good idea to keep Joseph in bed for a time, and we were not to put him under any sort of pressure or anxiety. This we decided to do. Tracey arrived about midday, and she came over to the flat to read to Joseph and also attend to some office detail with him. Joseph remained in bed. After Tracey's arrival, I was due to leave for a couple of hours.

Joseph seemed not to be himself, although quite aware of what was going on around him.

I returned at 4 p.m., and when everyone had left and I was on my own again with Joseph (Tracey returned to the Institute at 5 p.m.), the Dean of Gonville and Caius College arrived, and that pleased me because I hadn't thought about calling him in, and he was able to say a little prayer with Joseph in the study. He informed me that he would come the following morning for Joseph to take Holy Communion at 9 a.m., which I agreed would be fine, knowing this meant that I could put up the little table as an altar, so that Joseph could light a candle.

The nurse and Joseph's co-carer had decided to get Joseph up and sit him downstairs again. I thought this would be rather distressing for Joseph. His co-carer left after telling me that there were two nurses coming that evening to help me get Joseph into bed, which was a very good idea. But at 7 p.m., Joseph did not seem to be comfortable and seemed to become stressful, and, as I had no idea what time the nurses would arrive, I decided to put him to bed. I lifted him into his wheelchair, took him to his bedroom and lifted him onto his bed, sitting him up with lots of pillows and drinks. He had a sheepskin rug between his body and the sheet for extra comfort and warmth.

I sat talking to him for some time, and I thought it would be rather nice if I could arrange for someone to visit him, and thought of Elinor Shaffer, a very dear friend of Joseph's who lives near by, and I rang her. I found that she was in bed with a heavy cold, but despite this she did not hesitate to come round, indeed she arrived very quickly. I was able to talk with Joseph, and he would communicate by squeezing our hands, once for 'yes' and twice for 'no'.

After giving Joseph a drink, just before Elinor arrived, I put my arm around his shoulder and he held my hand very tightly and I said, 'Joseph – are you frightened?' 'Oh no,' was the reply in a weak voice. I said, 'I am very pleased to hear that, because this is a path we all must tread. There is no turn-back, but that at the end of this path, you will meet your lovely Dophi and your lovely Gwei-Djen.' Elinor arrived, and before going into Joseph's room, I explained to her that Joseph was not quite the same Joseph she had seen a few weeks before. She sat very close to him and he knew exactly what Elinor was talking about, as she described her recent lecture tour. From time to time I went in to them, and Joseph was quite awake, eyes wide open. Elinor had brought in a daffodil for him.

I took this opportunity to clean up a little and prepare for the night-time routine. About 8.30 p.m. the nurses arrived, and I explained that he was already in bed, but that I would appreciate their help to make him comfortable. The nurses agreed to give Elinor a little more time with Joseph, while I talked to them in the dining-room. At about 8.45 p.m. I went in to talk with Elinor to explain that the nurses had arrived, telling Joseph how lucky he was to have two beautiful young girls to make him comfortable, to which he gave that well-known big smile again.

Elinor decided that she would leave, but I persuaded her to wait in the study/lounge for a cup of tea, and she turned to say good-bye to Joseph, holding his proffered hand. I was holding his other hand, she said, 'Good-bye Joseph – I shall be in to see you soon.' With that he gave a slight sigh and he died. I looked at Elinor and said that he had just died, Elinor did not realize that this had happened. He went so peacefully; and the nurses had just entered the room. They, too, were quite surprised that there was no pain, and no fighting for breath – it was exactly 8.55 p.m. when Joseph died.

We all had a cup of tea, and I informed the doctor, who came to certify the death, after which Joseph was taken to the Chapel of Rest. Elinor and the two nurses left, as did the doctor; then the undertakers arrived, and I saw our dear Joseph leave the house for the last time. I then had the task of ringing around and I rang the Dean.

Here are some things that people would never know – Joseph would sing, he would tell a little joke, or I would tell him a joke which would make him laugh. I recall one occasion, 11.45 p.m. one night he was singing 'The Red Flag' so loudly, I said, 'Joseph, they will hear you at Caius', and he replied, 'I do hope so, dear boy!' and continued singing. It was a real pleasure to work with Joseph. When I returned from my week's break, he would take my hand into his and say, 'Welcome back.'

I look upon this four years nursing and caring for Joseph as the greatest privilege of my life. One day I hope to extend my hand to Joseph, when he will say, 'Welcome back' with that lovely warm smile. Until then, 'Good-bye, Joseph.'

# 3

## Honours
## and decorations

Emeritus Director, Needham Research Institute (East Asian
 History of Science Library), Cambridge (Director,
 1976–90)
Fellow of the Royal Society, 1941
Fellow of the British Academy, 1971
Master of Gonville and Caius College, 1966–76
Honorary Counsellor, UNESCO
Chairman, Ceylon Government University Policy
 Commission, 1958
President, International Union of History of Science, 1972–75
Foreign Member: National Academy of Sciences, USA
 American Academy of Arts and Sciences
 American Historical Association
 National Academy of China (Academia Sinica)
 Royal Danish Academy
Member: International Academies of History of Science and
 of Medicine
 Honorary Member of Yale Chapter of Sigma Xi
Honorary Fellow, UMIST
Honorary FRCP, 1984
Honorary D.Sc., Brussels, Norwich, Chinese University of
 Hong Kong
Honorary LL.D., Toronto and Salford
Honorary Litt.D.: Cambridge, Hong Kong, Newcastle-upon-
 Tyne, Hull, Chicago
 Wilmington, N.C., and Peradiniya, Ceylon; University of
 Surrey

Honorary Ph.D. Uppsala
Sir William Jones Medallist
Asiatic Society of Bengal, 1963
George Sarton Medallist
Society for History of Science, 1968
Leonardo da Vinci Medallist
Society for History of Technology, 1968
Dexter Award for History of Chemistry, 1979
Science Award (First Class), National Science Commission of
    China, 1984
International Science Policy Foundation Medal, 1987
Fukuoka Municipality Medal for Asian Culture, 1990
Order of the Brilliant Star, Third Class with sash (China)
Friendship Ambassador, 1990 (title conferred by Chinese
    People's Committee for Friendship with other Countries)
Foreign Academician, Academia Sinica, Beijing, 1994
UNESCO Einstein Gold Medal, 1994

# Publications

*Books by Joseph Needham, 1925–81*
*(Titles edited by Needham are preceded by an asterisk)*

*Science, Religion and Reality*. London, Sheldon Press, 1925; New York, Macmillan, 1925, 1928; New York, Braziller, 1955.

*Chart to Illustrate the History of Physiology and Biochemistry*. Cambridge, Cambridge University Press, 1926.

*Man a Machine*. London, Kegan Paul, 1927; New York, Norton, 1928.

*The Sceptical Biologist*. London, Chatto & Windus, 1929; New York, Norton, 1930.

*Materialism and Religion*. London, Benn, 1929.

*Chemical Embryology* (3 vols.). Cambridge, Cambridge University Press, 1931; New York, Hafner, 1963.

*The Great Amphibium*. London, SCM Press, 1931; New York, Scribner, 1932.

*A History of Embryology*. Cambridge, Cambridge University Press, 1934, 1959; New York, Abelard-Schuman, 1959; New York, Arno, 1975. Russian translation: Moscow, 1947.

*Background to Modern Science*. Cambridge, Cambridge University Press, 1935, 1940; New York, Macmillan, 1938; New York, Arno, 1975.

*Christianity and the Social Revolution*. London, Gollancz, 1935, 1937; New York, Scribner, 1936.

*Adventures before Birth*. London, Gollancz, 1936. (Translated from Jean Rostand.)

*Order and Life.* New Haven, Conn., Yale University Press, 1936; Cambridge, Cambridge University Press, 1936; paperback edition: Cambridge, Mass., MIT Press, 1968. Italian translation: Turin, Einaide, 1946.

\*\*Perspectives in Biochemistry* (F. G. Hopkins Presentation Volume). Cambridge, Cambridge University Press, 1937.

*Integrative Levels: A Revaluation of the Idea of Progress.* Oxford, Oxford University Press, 1937.

*The Levellers and the English Revolution* (under *nom de plume* Henry Holorenshaw). London, Gollancz, 1939; New York, Fertig, 1971. Translations: Russian, Moscow, 1947; Italian, Milan, Feltrinelli, 1957.

*The Nazi Attack on International Science.* London, Watts, 1941.

*Biochemistry and Morphogenesis.* Cambridge, Cambridge University Press, 1942, 1950, 1966 (with new introduction); New York, Macmillan, 1942.

\*\*The Teacher of Nations: Commemoration of Comenius.* Cambridge, Cambridge University Press, 1942.

\*\*Science in Soviet Russia.* London, Watts, 1942; New York, Arno, 1975.

*Time, The Refreshing River.* London, Allen & Unwin, 1943; New York, Macmillan, 1943.

*Chinese Science* (album of pictures taken during Second World War). London, Pilot Press, 1945.

*Science and Unesco.* Paris, UNESCO, 1946.

*History is on Our Side.* London, Allen & Unwin, 1946; New York, Macmillan, 1947.

*Science and Society in Ancient China.* London, Watts, 1947.

*Science Outpost* (with D. M. Needham). London, Pilot Press, 1954. Translations: Chinese, Shanghai, Chunghua, 1947, Taipei, Chunghua (in 2 vols.), 1952, 1955; Japanese, Tokyo, Heibonsha, 1986.

*Unesco Liaison.* Paris, UNESCO, 1946.

*Science and International Relations.* Oxford, Blackwell, 1949; New York, Thomas, 1949.

*Hopkins and Biochemistry* (F. G. Hopkins Memorial Volume). Cambridge, Heffer, 1949.

*Human Law and the Laws of Nature in China and the West.* Oxford, Oxford University Press, 1951.

*Science and Civilisation in China* (7 vols in 20 parts) (with many collaborators). Cambridge, Cambridge University Press,

1954. Translations: Japanese, Tokyo, Sushakusha, 1974; Chinese, Beijing, 1975, Taipei, 1971.

*Chinese Astronomy and the Jesuit Mission: An Encounter of Cultures.* London, China Society, 1958.

*The Development of Iron and Steel Technology in China* (with Wang Ling). London, Newcomen, 1958, 1964; Cambridge, Cambridge University Press, 1970.

*Heavenly Clockwork: The Great Astronomical Clocks of Mediaeval China* (with Wang Ling and Derek de S. Price). Cambridge, Cambridge University Press (for the Antiquarian Horological Society), 1960.

*Classical Chinese Contributions to Mechanical Engineering.* Newcastle, Kings College, 1961.

*Time and Eastern Man.* London, Royal Anthropological Institute, 1965.

*Within the Four Seas.* London, Allen & Unwin, 1969. Translations: Italian, Milan, Feltrinelli, 1975; Spanish, Mexico City, Siglo XXI, 1975.

*The Grand Titration.* London, Allen & Unwin, 1969; paperback edition: London, Allen & Unwin, 1979. Translations: French, Paris, Seuil, 1973, 1978; Italian, Milan, Mulino, 1973; Japanese, Tokyo, Hosei, 1975; Spanish, Madrid, Alianza, 1977.

\**The Chemistry of Life.* Cambridge, Cambridge University Press, 1970. Translations: Spanish, Mexico City, Cultura Económica, 1974; Japanese, Tokyo, 1978.

*Clerks and Craftsmen in China and the West* (with several collaborators). Cambridge, Cambridge University Press, 1970. Translations: Japanese (2 vols.), Tokyo, 1974; Spanish, Mexico City, Siglo XXI, 1978.

*The Chinese Contribution to the World.* Tokyo, Kinseido, 1973.

Von der Vielfalt der Traditionen im modernen China (Festschrift). *Die Waage*, Vol. 5. (Edited by M. Teich and R. Young.)

*Changing Perspectives in the History of Science* (Festschrift). London, Heinemann, 1973. (Edited by S. Nakayama and N. Sivin.)

*Chinese Science: Explorations of an Ancient Tradition.* Cambridge, Mass., MIT Press, 1973.

*La tradition scientifique chinoise.* Paris, Hermann, 1974.

The Nature of Chinese Society – A Technical Interpretation (with Huang Jen-Yü). *Journal of Oriental Studies.* 1974, Vol. 12,

No. 1/2, p. 1 (Republished in *East and West*, Vol. 24 (new series) 1974.)

*Moulds of Understanding.* London, Allen & Unwin, 1976. Spanish translation: Barcelona, Critica, 1978.

*Wissenschaftliche Universalismus.* Frankfurt/Main, Suhrkamp, 1977; paperback edition, 1979.

*Chinas Bedeutung für die Zukunft der westlichen West.* Cologne, Deutsche-China Gesellschaft, 1977.

*The Shorter Science and Civilisation in China* (with Colin Ronan). Cambridge University Press, Cambridge, 1978. Parallel Chinese abridgement: Taipei, Com Press, 1972.

*Three Masks of the Tao.* London, Teilhard Centre, 1979.

*Celestial Lancets: A History and Rationale of Acupuncture and Moxa* (with Lu Gwei-Djen). Cambridge, Cambridge University Press, 1980.

*Science in Traditional China: A Comparative Perspective.* Hong Kong, Chinese University Press, 1981.

## *S e r m o n s*

List of Joseph Needham's sermons 1963–87
All sermons were given at Thaxted Church unless otherwise stated.

| *Occasion*/place | Date | *Topic*/occasion |
|---|---|---|
| *Epiphany 4* | 29 January 1984 | *Torture* |
| *Trinity 15* | 11 September 1983 | *Martyrs* |
| *Trinity 2* | 12 June 1983 | *Ethics of Economics and Science* |
| *Easter 1* | 10 April 1983 | *Mammon* |
| *Advent 2* | 5 December 1982 | *Deferring and Democracy* |
| *Quinquagesima* | 13 February 1983 | *Against 'Holy War' Idea* |
| *Trinity 17* | 3 October 1982 | *History of the Church and Science Fiction* |
| *Trinity 9* | 8 August 1982 | *'Try the Spirits'* |
| *Easter 5* | 16 May 1982 | *The 'Just War'* |
| *Lent 5* | 28 March 1982 | *Religion and Superstition* |
| *Epiphany 3* | 24 January 1982 | *Magi in White Overalls* |
| *Trinity 22* | 15 November 1981 | *Shintoism, Japanese Confucianism and the Cosmic Christ* |
| *Trinity 8* | 9 August 1981 | *The Spirit of Evil in Things Heavenly* |

| Occasion/place | Date | Topic/occasion |
|---|---|---|
| *Trinity Sunday* | 14 June 1981 | *The Social Aspect of Trinity Sunday* |
| *Easter 1* | 26 April 1981 | *Resurrection and Progress* |
| *Septuagesima Sunday* | 15 February 1981 | *Robots and Unemployment* |
| *Advent 3* | 14 December 1980 | *Messianic Prophecy* |
| *Friar Jack's Requiem* | 26 November 1980 | *Address* |
| *Trinity 18* | 5 October 1980 | *Fundamentalism* |
| *Hiroshima Day* | 6 August 1980 | Address at planting of commemorative tree |
| *Trinity 2* | 15 June 1980 | *Jealousy* |
| *Easter 2* | 20 April 1980 | *Shepherds and Conflict* |
| *Advent 2* | 9 December 1979 | *Gnosticism and Evil* |
| *Trinity 11* | August 1979 | *Original Sin* |
| *Trinity 3* | 1 July 1979 | *Humility* |
| *Trinity 1* | 1979 | *Love* |
| *Easter 5* | — | *Political Prisoners and Torture* |
| *Easter 2* | 29 April 1979 | *Sacrifice* |
| Caius Chapel | 22 January 1961 | Address on Religion East and West |
| *Union Society Speech* | 9 May 1961 | |
| Girton College Chapel | 5 February 1978 | *The Church in China* |
| College Chapel, Oxford University Sermon | 1 May 1977 | *Three Masks of the Tao* |
| Caius Chapel | Whitsun 1976 | *Love* |
| *Perse Sermon* | 14 December 1973 | *Place of Women in the Church* |
| — | 22 September 1976 | Requiem for Tom Driberg |
| Caius Chapel | 23 February 1975 | Norman Bethune Address |
| *Advent Sunday* | 1973 | Denver Sermon |
| Caius Chapel | 9 March 1973 | Requiem for Bruno Renner |
| Caius Chapel | 4 June 1972 | *'The Inner Life'* |
| Trinity Hall Chapel | 10 May 1970 | *Science and Christianity* |
| Caius Chapel | 20 February 1970 | *Time* |
| Perse School and Newport School | 20 October 1969 | Speech Day |
| Caius Chapel | 11 May 1969 | *Christianity and Marxism* |

| *Occasion*/place | Date | *Topic*/occasion |
|---|---|---|
| Caius Chapel | 25 February 1968 | BMV Sermon |
| Caius Chapel | 27 October 1963 | Commemoration of Benefactors |
| Caius Chapel | 31 October 1965 | Commemoration of Benefactors |
| Caius Chapel | 28 October 1973 | Commemoration of Benefactors |
| *Trinity 4* | 12 July 1987 | *Greed and Capitalism* |
| *Easter 2* | 13 May 1987 | *Prophecy* |
| Grantchester | 6 January 1987 | Margaret Hatfield's funeral |
| *Trinity 14* | 31 August 1986 | *Sin and Original Sin* |
| *Trinity 6* | 6 July 1986 | *Redemption* |
| *Trinity Sunday* | 25 May 1986 | *Doctrine of the Holy Trinity* |
| — | 22 July 1985 | Lettice Ramsay's cremation |
| Trinity 5 | 7 July 1985 | *Violence Everywhere Today* |
| *Whit Sunday* | 26 May 1985 | *International Language and Science, the Universe* |
| *Sexagesima* | 10 February 1985 | *Miracles and Symbolism* |
| *Christmas 1* | 30 December 1984 | *'Revelations', Science Fiction* |
| *Hiroshima Day* | 5 August 1984 | Ullambana Ceremony |
| *Trinity 3* | 8 August 1984 | *Ceremonial Due Order, Renunciation, Obsession* |
| *Easter 5* | 27 May 1984 | *Death* |
| *Lent 2* | 18 March 1984 | *Women* |
| Faculty of Divinity | 8 November 1972 | *Christian Hope and Social Evolution: Thien Hsia Ta Thung and Regnum Dei* |

# Index